T0366553

# SEA MAMMALS

Published by Princeton University Press
41 William Street, Princeton, New Jersey 08540
99 Banbury Road, Oxford OX2 6JX
press.princeton.edu



Library of Congress Control Number 2022950268
ISBN 978-0-691-23664-3
Ebook ISBN 978-0-691-24338-2

Typeset in Aviano Flare and Adobe Caslon Pro

Printed and bound in China
10 9 8 7 6 5 4 3 2 1

British Library Cataloging-in-Publication Data is available

This book was conceived, designed, and produced by UniPress Books Limited
Publisher: Nigel Browning
Commissioning editor: Kate Shanahan
Project manager: Kate Duffy
Art direction and design: Alexandre Coco
Illustrator: Bob Nicholls
Picture researcher: Alison Stevens

COVER IMAGES:
(Front cover)
Flip Nicklin / Minden Pictures / naturepl.com

(Back cover)
Bob Nicholls, Paleocreations

## 4
## BEHAVIOR



## 5
## ECOLOGY AND CONSERVATION

# INTRODUCTION

Today, sea mammals include some 137 living species (five extinct) worldwide. They comprise a diverse assemblage of at least seven distinct evolutionary lineages that evolved on land and independently returned to the sea, spending most of their time in water.

The majority of sea mammals are cetaceans (whales, dolphins, and porpoises). The name Cetacea is from the Greek *cetos*, meaning "whale." The 93 species (one extinct) currently grouped in the Cetacea are divided into two major groups: toothed whales or Odontoceti and baleen whales or Mysticeti. Toothed whales are considerably more diverse with ten families, 33 genera, and 78 species (one extinct). Toothed whales include sperm whales (Physeteridae), pygmy sperm whales (Kogiidae), beaked whales (Ziphiidae), oceanic dolphins (Delphinidae), river dolphins (Platanistidae, Lipotidae, Pontoporiidae, and Iniidae), narwhal and beluga (Monodontidae), and porpoises (Phocoenidae). Baleen whales include four families, six genera, and 15 species of right whales and bowheads (Balaenidae), pygmy right whale (Neobalaenidae), gray whale (Eschrichtiidae), and blue, fin, sei, Bryde's, humpback, minke, Omura's whale, and the recently described Rice's whale (Balaenopteridae).

Pinnipeds are named from the Latin *pinna* and *pedis*, meaning "feather footed," referring to their paddle-like fore and hind limbs. Pinnipeds include 35 species (two extinct) of earless seals (Phocidae), fur seals and sea lions (Otariidae), and walruses (Odobenidae). Phocid seals are the most diverse with 19 species (one extinct): 15 species (one extinct) of fur seals and sea lions, and a single living walrus species.

Sirenians or sea cows are named from mermaids of Greek mythology (see page 64). There are five species of sirenians (one extinct), three species of manatee (Trichechidae), and two dugongs (Dugongidae). Sirenians are unique among living marine mammals in having an herbivorous diet, feeding almost entirely on aquatic plants.

In addition, there is one bear species (Ursidae), the polar bear, two species of sea and marine otters, and one extinct species of sea mink (Mustelidae).

Several smaller groups are entirely extinct: hippopotamus-like desmostylians, aquatic sloths (*Thalassocnus*), and the oyster bear *Kolponomos*.

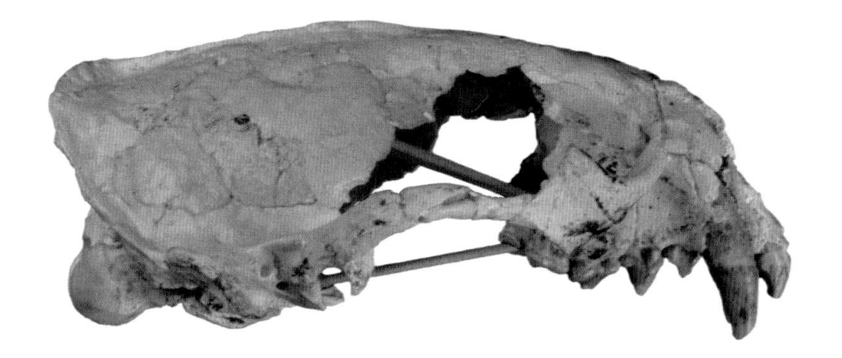

The skull and lower jaw of the extinct giant otter *Enhydritherium terranovae* showing teeth with thick enamel and low rounded cusps.

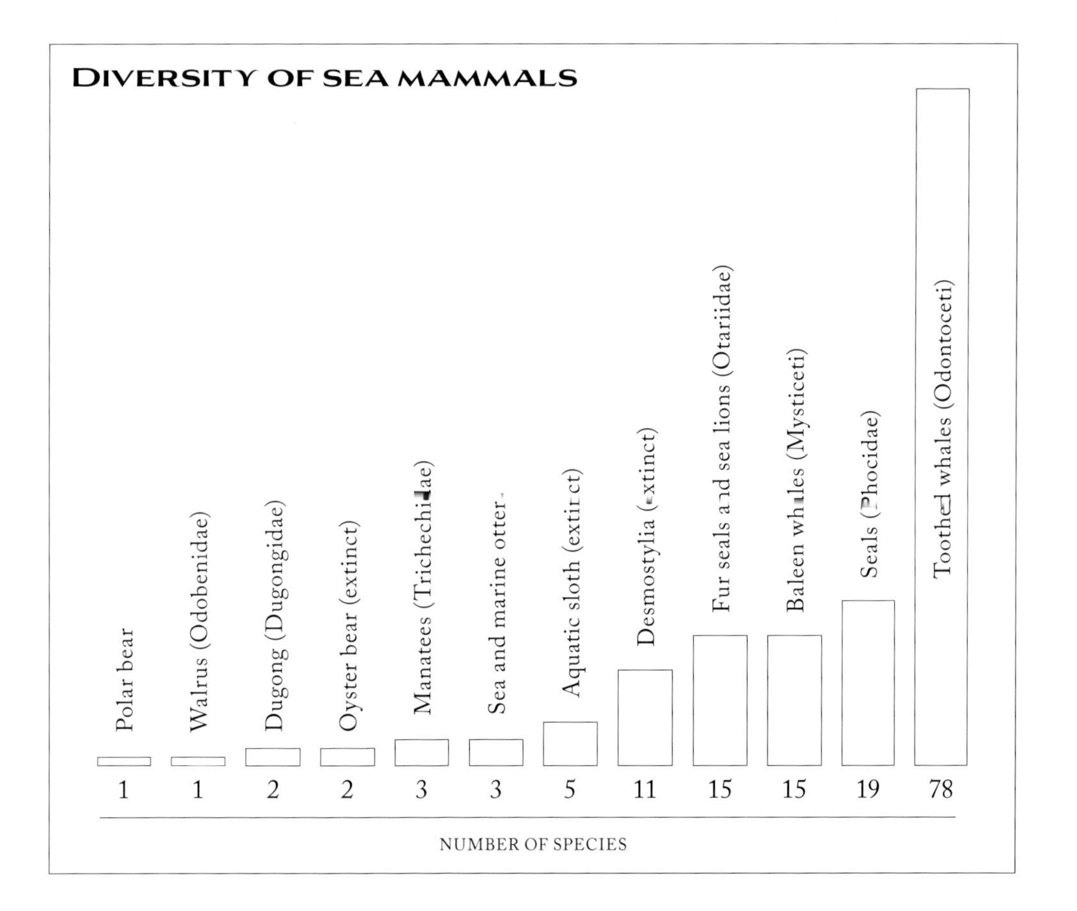

# Evolutionary history

The major groups of sea mammals have separate origins from different, land-dwelling ancestors at various times in the past. Cetaceans evolved a little more than 50 million years ago (mya) in the shallow Tethys Sea, which extended from the Pacific to the present-day Mediterranean Sea. Cetaceans evolved from within even-toed, hoofed mammals (artiodactylans, e.g., hippos, deer, and sheep) and for this reason are grouped in the Cetartiodactyla. Pinnipeds are aquatic carnivores and evolved about 27 mya from arctoid carnivorans, for example, ursids (bears), procyonids (racoons and allies), and mustelids (otters, weasels, and kin), in the North Pacific. Sirenians, like cetaceans, have a long evolutionary history, extending back 50 mya in the Tethys Sea, but they differ in having evolved from elephants, hyraxes, and kin (Paenungulata). The only extinct order of marine mammals, the Desmostylia, related to odd-toed hoofed mammals, for example, extinct relatives of horses, tapirs, and rhinos, evolved in the North Pacific where they were confined from 33–10 mya. Marine and sea otters evolved from river otters several million years ago and the polar bear evolved from brown bears less than 1 mya. Aquatic sloths comprise five species in a single genus *Thalassocnus* and share a common ancestry with ground sloths, which evolved in western South America approximately 8 mya where they remained until their extinction 1.5 mya (see page 24).

## Geologic and climatic events

| Time (mya) | Epoch | | Events |
|---|---|---|---|
| 0 | Pleistocene | | |
| 5 | Pliocene | Late | Bering Straits open; Bering and Chukchi Seas created |
| | | Early | Panamanian seaway closed |
| 10 | Miocene | Late | Bering Straits and Tethys Sea closed and reduced to a series of lakes |
| 15 | | Middle | Panamanian seaway and Bering Straits open<br>Australia continued to drift north, closing the Indo-Pacific seaway between Australia and Asia, restricting equatorial circulation of Indian and Pacific Oceans |
| 20 | | Early | |
| 25 | Oligocene | Late | Antarctica separated from South America, initiating east-west flow of the Antarctic Circumpolar Current |
| 30<br>35 | | Early | Australia moved north away from Antarctica, opening part of the Southern Ocean, resulting in major climatic changes as colder, drier weather replaced a wet and warm world |
| 40 | Eocene | Late | |
| 45<br>50 | | Middle | As a result of India colliding with Asia, much of the Tethys seaway closed, resulting in climatic cooling and ocean restructuring<br>Shallow, extensive Tethys seaway opens from Pacific to Mediterranean |
| 55 | | Early | |

An adult leopard seal (*Hydrurga leptonx*) located on the Antarctic Peninsula. The spotting on the neck, throat, and chest are distinguishing features of this species of seal.

## ADAPTATIONS FOR LIFE IN WATER

The secondary adaptation of sea mammals to life in water required various anatomical specializations compared to their land-dwelling ancestors. For the major lineages this includes the evolution of relatively large, streamlined bodies with reduced appendages (for example, small or no external ears) and limbs transformed into flippers. Pinnipeds, marine otters, and the polar bear spend time on the land and are able to "walk" with modified limbs or flippers. Cetaceans and sirenians spend their entire lives in water and they have lost their hind limbs, instead employing their tails for propulsion. For some marine mammals, such as fur seals, sea lions, and sea and marine otters, a layer of fur keeps them warm. Cetaceans and sirenians have lost fur, with cetaceans possessing a thick layer of subcutaneous fat or blubber for temperature regulation, energy storage, and streamlining. Cetaceans and sirenians retain only a scattering of hairs on the body surface that have a sensory role. Some pinnipeds (for example, Weddell and elephant seals) and a number of toothed whales (for example, sperm and beaked whales) are incredible diving machines capable of prolonged and deep dives on a single breath of air and have specializations such as flexible ribs that allow the lungs to collapse, a slowed heart rate, reduced oxygen consumption rates, and blood that is pumped only to essential organs.

The endangered Ganges river dolphin (*Platanista gangetica*), located in the Brahmaputra River, India. Distinctive features of this species include a long upturned beak, protruding teeth, and tiny eyes.

## Behavior

Most toothed whales live in large social groups called schools, pods, or units. Baleen whale social groups are typically small. This probably reflects their need to feed more or less individually given their large daily food requirements. Many pinnipeds, especially those breeding on land (for example, elephant seals) are highly polygynous with a few males mating with a large number of females. Ice breeding pinnipeds tend to be solitary, since ice is an unstable habitat, only occasionally forming small mating groups. Sirenians, solitary for most of the year, are promiscuous, mating with multiple partners, and in some locations form mating herds. Marine otters and polar bears mate on land and have polygynous breeding systems. Sea otters mate in the water and are also polygynous.

## Food and Feeding

Most pinnipeds are pierce feeders using their teeth to capture and pull prey in the mouth, swallowing prey whole, and feeding on various fish, with several being specialized squid (for example, elephant seals) or mollusk feeders (for example, walrus). Mysticetes use baleen plates that hang down from the upper jaw as sieves to filter feed concentrations of fish and small invertebrates such as krill. Toothed whales feed on various fish and invertebrates, obtaining prey singly, aided by their teeth and echolocation (high-frequency sound) abilities. Sirenians use their fleshy, mobile lips to graze on aquatic plants.

Extinct desmostylians and aquatic sloths likely also had fleshy lips that together with their high crowned teeth were used to consume seagrasses. The rounded, heavy molars of sea otters are used for crushing bottom dwelling invertebrates including abalones, sea urchins, and rock crabs.

## Life History

Sea mammals typically are very long lived (many live 80–90 years), with the bowhead reportedly living at least 200 years. Most marine mammal give birth to a single offspring (marine otters and the polar bear are exceptions) and intensively nurse their young with high-fat milk. Pinnipeds, otters, and the polar bear mate and give birth on land, or in water. Cetaceans and sirenians conduct reproductive activities entirely in the water.

## Distribution and Habitat

Most marine mammals occupy the world's ocean basins and have specific habitat requirements, such as shallow versus deep water, although a few range broadly around the world. Manatees and river dolphins inhabit estuarine or fresh water. Some marine mammals such as baleen whales, sperm whales, and some pinnipeds, such as elephant seals, undergo long-distance annual migrations to feed in productive high latitudes in the summer, and they spend the winter fasting and give birth in lower latitudes.

## Ecology and Conservation

Sea mammals have been extraordinarily successful over the past 50 mya and they remain ecologically important in the twenty-first century. Broad aspects of their biology are correlated with their ecological role and their response to the changing distribution of resources, such as food and territory. They have occupied apex roles in food webs, feeding at both the base of the food chain, for example, sirenians, and at higher trophic levels, such as pinnipeds, whales, polar bears, sea and marine otters. The survival of marine mammals has depended on their ability to adapt to global climate change and other threats.

Several marine mammals have been hunted to extinction by humans in historic times, such as the Yangtze river dolphin (*Lipotes vexillifer*), Japanese sea lion (*Zalophus japonicus*), Caribbean monk seal (*Neomonachus tropicalis*), and Steller's sea cow (*Hydrodamalis gigas*). Some sea mammals are on the brink of extinction, including the vaquita (*Phocoena sinus*), the North Atlantic right whale (*Eubalaena glacialis*), and the Mediterranean monk seal (*Monachus monachus*). However, there are also exciting discoveries of new species, such as Sato's beaked whale (*Berardius minimus*), Omura's whale (*Balaenoptera omurai*), and Rice's whale (*Balaenoptera ricei*).

Sea mammals have been exploited for centuries by humans on all continents. Among various threats are hunting, incidental mortality in fishing nets, pollution, human generated noise (shipping and seismic survey), and climate change. The commercial hunting of large whales (baleen and sperm whales) began in the North Atlantic in the 1600s and spread around the world, continuing until international regulations banned whaling in the mid-1980s. Efforts to conserve sea mammals began in the twentieth century. In recent decades, the hunting of smaller whales and dolphins, though significantly reduced, still occurs.

A subspecies of the West Indian Manatee (*Trichechus manatus latirostris*) has its skin cleaned by blue gill fish in Three Sisters Springs, Crystal River, Florida.

Some pinnipeds, such as northern fur seals, elephant, and monk seals, were heavily exploited for their oil and pelts and went through severe population declines, including the Caribbean monk seal, which was hunted to extinction in the mid-twentieth century.

Sirenians have also been hunted in the past. Current threats include boat collisions and red tide poisoning for manatees and habitat degradation for dugongs. Sea otters have been subjected to population fluctuations, victims of changing food web dynamics and oil spills. The biggest threat to polar bears, considered the poster child of climate change, is loss of sea ice habitat.

## SCOPE AND ORGANIZATION

This book highlights more than 50 iconic sea mammals, including fossil and living species of whales, pinnipeds, sea cows, sea and marine otters, and the polar bear, and tells their unique stories. Species were selected for intriguing stories of their evolution: their origin, relatives and relationships, and when they lived; and for how they were discovered and collected. The biology of selected marine mammals explores their unique anatomical adaptations, such as tusks, body sizes, flippers, flukes, and melon. Aspects of the fascinating behavior of sea mammals are discussed, for example, diving, feeding, and locomotion. Species were selected to show how ecology and conservation relate to life histories and status, threats, and actions. Locator maps show the distribution of species. And finally, quick facts detail diet, habitat, life history, and conservation status, lengths, and weights; please note measurements are approximate.

# 1
# EVOLUTION

A historical framework provides highlights of the past lives of sea mammal species, when and where they originated, how they evolved, and what led to their extinction.

Among fossil whales, tales are told of an early whale that had legs and evolved on land in India and Pakistan. Several fossils show interesting adaptations, such as the bizarre walrus-like whale known from the Southern Hemisphere that evolved tusks, which may have aided in feeding; and the skim-feeding porpoise from California that used its unusual, elongated lower jaws to probe sediments for prey. An extinct baleen whale from the North Pacific is hypothesized to represent a transitional form. It possessed both baleen and teeth, differing from its ancestors that only had teeth and pursued single prey items. Modern adult mysticete whales, which replaced teeth with keratinized baleen plates, bulk filter feed on the ocean's tiniest animals. The fossil giant sperm whale from Peru likely used its large teeth to tear the flesh of other large mammals, unlike the modern squid-eating sperm whale. The evolution of another toothed whale from the southeastern US provides important clues about the origin of high-frequency hearing earlier than previously believed.

The earliest pinniped probably evolved in fresh water in the Arctic and aided by its webbed feet spent more time on land

PAGE 14
A skull and partial skeleton showing the vertebrae and ribs of an early fossil whale (*Basilosaurus isis*) was discovered at the UNESCO World Heritage Site of Wadi Al-Hitan, in Egypt.

than modern seals do. A fossil walrus from California and Baja California uniquely evolved only tusks that were probably used in social interactions.

Sea cows also evolved on land, walking on all four legs; the earliest representatives of the group are known from Africa. An extinct sea cow relative from the North Pacific provides evidence for the evolution of this lineage from an odd-toed common ancestor in Asia, and subsequent dispersal to the North Pacific where the group remained.

Adding to the rich, past diversity of marine mammals are marine sloths from the Southern Hemisphere, which specialized for feeding on aquatic plants; and the extinct "oyster bear" from the Pacific Northwest that likely crushed hard-shelled mollusks with a powerful bite. A similar diet is also suggested for a Southern Hemisphere fossil seal that had blunt, robust teeth. A giant fossil otter from the US and Mexico suggests that sea otters evolved in fresh water and dispersed along river systems to coastal environments.

# SKIM-FEEDING PORPOISE

## The skim-feeding porpoise is an extinct species from southern California; it may have used its unusually long lower jaw to skim the seafloor for food.

The skim-feeding porpoise (*Semirostrum ceruttii*) has a lower jaw extending beyond the upper jaw, making it the first mammal known to have a huge underbite. Paleontologist Rachel Racicot's first clue about the function of the lower jaw came to her when bird watching. She noticed that the porpoise jaw resembled that of the black skimmer, a seabird with an underbite that flies low over the water with its partially submerged lower jaw skimming for prey. The CT scans she analyzed showed that the porpoise's jaw had long canals that served as conduits for nerves and blood vessels like those found in black skimmers and the halfbeak fish, which suggested a similar function in the porpoise. Racicot also noticed that *Semirostrum ceruttii* had small optic canals, which suggests that it may not have used its eyesight as much as modern porpoises. While some modern porpoises and dolphins locate prey near the seafloor, they are not specialized for feeding there. Modern porpoises feed throughout the water column preferring bottom-dwelling and mid-water prey, including schooling fish, squid, and octopus. The difference is that during the Pliocene era there was an increase in species diversity that may have resulted in more species feeding on prey on the sea bottom. This interpretation is supported by the skim-feeding porpoise's well-developed neck muscles and robust, unfused neck vertebrae that permitted increased mobility and maneuverability of the head, consistent with an animal probing the seafloor. Known as benthic probing, this may have evolved in low light or conditions unsuitable for the use of echolocation (see page 44).

Porpoises are among the smallest of modern whales. They range in size from the vaquita porpoise (see page 170) at less than 5 feet long (1.4 m) to the Dall's porpoise (see page 152) at more than 7 feet long (2.3 m). Depending on the taxonomy followed, there are six or seven extant species in the porpoise family Phocoenidae. Porpoises have an excellent fossil record that goes back to the late Miocene and Pliocene periods in the North Pacific and the North and South Atlantic.

SIZE

Males & females
Length: 6 ft 6 in (2 m)
Weight: 440 lb (200 kg)

DIET

Bottom feeding

HABITAT

Coastal marine

AGE

5.3–1.6 mya

The extinct skim-feeding porpoise had a large underbite of the lower jaw, which was partially submerged when probing the sediments for prey.

The skim-feeding porpoise is one of 15 species of fossil porpoise. The majority of fossil phocoenids are from the North Pacific, supporting a Pacific origin for the group as does the Pacific occurrence of related delphinoids. Porpoises are most closely related to dolphins, belugas, and narwhals. They can be distinguished from dolphins by their blunt snout, as opposed to the beaked and elongated snout of dolphins.

# Walking Whale

## The extinct walking whale, *Ambulocetus natans,* had well-developed legs and feet, and remarkably it could both walk on land and swim proficiently.

When it was discovered in Pakistan some 30 years ago, *Ambulocetus natans* (meaning walking whale that swims) was the only whale known to have legs. *Ambulocetus*, the best-known genus in the family Ambulocetidae, was described by paleontologist Hans Thewissen and colleagues in 1994 from sediments found in Indo-Pakistan, which were deposited in the warm, coastal waters of the Tethys Sea.

The walking whale resembled a crocodile with an extended snouted skull, long, flexible neck, short forelimbs, and a powerful tail that lacked the tail flukes present in living whales. The well-developed hind limb, webbed feet, and completely fused sacrum indicate that it was able to support its body weight on land, although it was probably slow and awkward. It was the size of a Northern sea lion (see page 196). Limb proportions were similar to those of a river otter, and it has been suggested that they swam in a similar manner using their hind limbs and tail, but differing from otters in that for the walking whale, the feet, not the tail, provided the main propulsion.

Like crocodiles, walking whales were ambush predators pursuing prey on land or in water. The long, sharp teeth and wear surfaces of this species are similar to other archaic whales, and suggest that they fed on fish and other animals that ventured into shallow water.

Features of the skull and jaws indicate the walking whale could hear underwater sound. Although the ear bones are not well preserved, based on fossil relatives *Ambulocetus* may have retained an ability to hear airborne sound. The eyes of the walking whale were placed high on the head facing upward. A similar position occurs in the hippopotamus and suggests that this early whale swam submerged, except for its eyes. The upward facing eyes would have helped locate prey swimming above the whale.

Fossils of *Ambulocetus* were recovered from sediments that probably comprised an ancient estuary, which contained a mixture of both marine and fresh water based on the isotopes of oxygen in its bones. Since fresh water and salt water have different ratios of oxygen isotopes, we can determine the type of water an animal drank by studying the isotopes that were incorporated into its bones and teeth as it grew. The isotopes show that *Ambulocetus* likely drank both fresh water and salt water, which fits well with the idea that these animals lived in estuaries or bays.

SIZE

Males & females
Length: 10 ft (3 m)
Weight: 310 lb–518 lb
(142 kg–235 kg)

DIET

Fish, ambush predator

HABITAT

Estuaries or bays

AGE

48 mya

The anatomy of the extinct walking whale indicates that it could swim using its powerful tail and support its body and move slowly on land.

Ambulocetids include three genera: *Ambulocetus*, *Gandakasia*, and *Himalayacetus* from the early to middle Eocene (ca. 48–46 mya) of India and Pakistan. *Gandakasia* was the first ambulocetid to be described in 1958 and is only known by a few teeth. *Himalayacetus*, based on a lower jaw, and named for its discovery in the Indian Himalayas, was originally described as a pakicetid and thought to be the oldest whale at 53.5 mya. It now appears that dating was based on associated fossils that washed in from older layers. Based on later systematic work, *Himalayacetus* was assigned to the ambulocetids. The Ambulocetidae are one of six families of archaic whales called archaeocetes known from early to middle Eocene (52–42 mya) sediments in Africa and North America, but are best known from Pakistan and India.

# Walrus-Like Whale

## The walrus-like whale is a bizarre extinct whale discovered in South America. It was found to have a face resembling a walrus, and possessed two tusks that may have stabilized the head to facilitate feeding.

An extinct relative of the beluga and narwhal, the walrus-like whale *Odobenocetops* is known by two species, *O. leptodon* and *O. peruvianus*, from the late Miocene-early Pliocene period of Peru and Chile. The name comes from the Greek *odon*, "tooth," and *baino*, "walk," and *ops*, "like," meaning "whale that seems to walk on its teeth."

The walrus-like whale is distinguished from all other whales in possessing no teeth except for tusks positioned below a short, rounded snout. The tusks are elongate, directed backward and sexually dimorphic; the right tusk of males is larger than the left, reaching about 3 feet (1 m) long, while the left one is shorter at ca. 10 inches (25 cm). This contrasts with the small tusks of females that has the right tusk slightly larger than the left. The arched roof of the mouth or palate, together with reduced teeth in the upper jaw, a highly vascularized blunt snout, and a face with strong muscle insertion areas suggest the presence of a powerful upper lip. These features are all associated with suction feeding in which the mouth works like a vacuum pump. Similar to the modern walrus, the walrus-like whale fed on benthic invertebrates, such as mollusks and crustaceans. The rich blood vessels to the upper lip of the walrus-like whale suggests the presence of vibrissae or whiskers, which would have aided in the search for prey. The facial region of the skull is not concave and the nose is positioned far forward in the walrus-like whale. Together these characteristics indicate that the melon—fatty tissue in the forehead of all toothed whales that acts as a sound lens to focus sound—was vestigial or absent; and it is likely that high-frequency hearing or echolocation (see page 44) was either absent or not well-developed in this species. Vision, however, was important as evidenced from its relatively large eyes that were oriented upward indicating binocular vision, again resembling the walrus, rather than to the sides of the head as in most dolphins.

In view of the exceptional specializations of *Odobenocetops*, it was referred to a new odontocete family, the Odobenocetopsidae. Its resemblance to the walrus in terms of cranial anatomy and inferred feeding habits is an unprecedented example of convergence in whales and in walruses.

Several hypotheses have been proposed for the tusks. They may have had a similar function to the tusks of walrus, which have been found to have a social

SIZE

Males & females
Length: 10 ft–13 ft
(3 m–4 m)
Weight: 330 lb–1,300 lb
(150 kg–600 kg)

DIET

Benthic invertebrates; suction feeder

HABITAT

Coastal marine

AGE

4–3 mya

The extinct walrus-like whale is shown in its swimming and feeding position. A distinguishing feature of this species is the longer tusk on the right-hand side of the head.

role and purpose in attracting mates or in competition for mates (see page 108). Paleontologist Christian de Muizon and colleagues argue instead that the tusks evolved as a sort of "sled runner," orientation guides that would keep the head at the optimal angle to the seafloor when feeding. This is supported by the presence of large, well-developed articular surfaces connecting the head to the vertebral column, suggesting that it had strong, flexible neck musculature and swam with its head bent ventrally.

In the Southern Hemisphere, it appears that the walrus-like whale occupied an ecological niche otherwise unknown among toothed whales that was similar to that of the odobenine walruses, which were widespread in the Northern Hemisphere during the Pliocene.

# Aquatic Sloth

## Today sloths live in trees in tropical rain forests, but in the past a lineage of sloth was adapted to living in the shallow coastal waters of South America.

Aquatic sloths (*Thalassocnus* spp.) are in the family Megatheriidae, which includes the extinct giant ground sloth *Megatherium*. Currently, six species of sloth exist, and they are all relatively small and spend most of their lives hanging upside down in the trees of tropical rain forests in Central and South America. Since the late Eocene (35 mya) a greater diversity of sloths lived not just in trees but also on the ground and in water, ranging in size from the elephant-sized ground sloth to the pig-sized aquatic sloth. One of the more extraordinarily marine-adapted mammals, is a lineage of "ground sloths" belonging to the Xenarthra—a group that includes, in addition to sloths, anteaters, and armadillos.

*Thalassocnus natans* is one of five species of the extinct marine sloth, living from 8 to 1.5 mya, which have been described from what were coastal deserts in Peru and Chile from the Pisco Formation, which comprises a rich marine vertebrate fauna. These aquatic sloths lived alongside whales, pinnipeds, seabirds, sharks, and crocodylians. Features of their jaws suggest that they were herbivores and grazed on seaweed and seagrasses in shallow coastal waters. Analysis of CT images of the skull show a well-developed olfactory system, suggesting that they had an increased sense of smell, comparable to that of living sloths.

The ribs and limb bones of the later diverging species of *Thalassocnus* are thick and dense, which suggests an adaptation to reduce buoyancy (think diver's weight belts) helping them to sink to the seafloor to feed. Other aquatic mammals, such as otters, sea cows, and the platypus, tend to have thickened, dense bones compared to their land-dwelling relatives. For aquatic sloths this increase in bone density was also seen in skull bones, not seen in other aquatic mammals. Although there appears no functional advantage, it has been hypothesized as a correlation with or consequence of a whole-body increase in bone density. Systemic bone structure alteration, formerly known exclusively as a physiological adjustment (e.g., exercise related), was here shown to have been retained as an evolutionary by-product of their adaptation to shallow diving.

Aquatic sloths may have used their powerful claws to anchor themselves to the seafloor to facilitate feeding. The long tail was likely used for balance and diving rather than propulsion. Aquatic sloths were not undulating their spines up and down to swim like early whales. Instead, they bounded along the seafloor

SIZE

Males & females
Length: 8 ft (2.50 m)
Weight: Unknown

DIET

Seagrasses

HABITAT

Coastal marine

AGE

7.2–2.5 mya

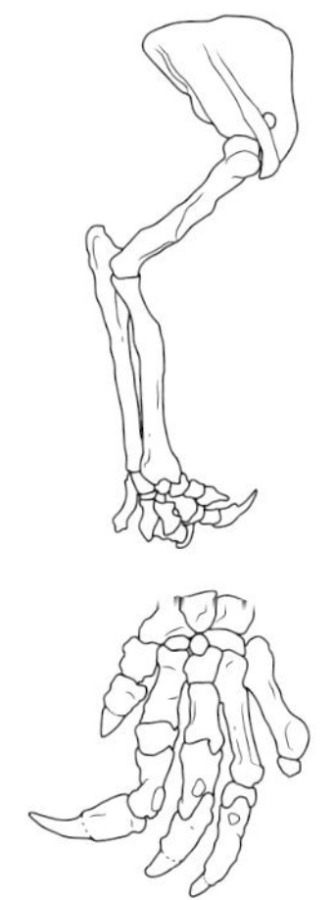

The extinct aquatic sloth had a long tail used for balance. The powerful forelimbs had large claws which were likely used as anchors in bottom walking and to uproot seagrasses.

employing bottom walking, pushing off the seafloor from point to point, much like modern hippos.

Aquatic sloths originated on land and gradually transitioned to life underwater. Comparison of the bones of dozens of skeletons from the oldest to youngest marine sloths, *T. antiquus* (8–7 mya), *T. natans* (6 mya), *T. littoralis* (5 mya), *T. carolomartini* (4–3 mya), and *T. yaucensis* (3–1.5 mya), found in different horizons of the Pisco Formation, reveal that sloths became increasingly more aquatic before becoming extinct. This scenario is supported by the recent discovery of *Thalassocnus* sp. from continental deposits in eastern Argentina. The relatively basal position of this new fossil shares the lowest semiaquatic capacity of species in the genus from Peru and Chile. Aquatic sloths may have exhibited sexual dimorphism; the skulls of individuals of some species show a size difference. They became extinct at the end of the Pliocene around the time the Isthmus of Panama closed and the Pacific Ocean was cut off from what is now the Caribbean Sea. As a result, the waters off South America became much colder than they had been, killing the seagrasses on which the sloths fed. Researchers think that aquatic sloths became extinct either due to the loss of their main food sources or because they were unable to tolerate cold water.

# Toothed Baleen Whale

## The teeth of modern baleen whales have been replaced by keratinous structures called baleen. Some fossil baleen whales had both teeth and baleen.

Archaic baleen whales include both toothed and toothless species. *Aetiocetus weltoni* is the best-known species of toothed baleen whale. Fossils have been found from rocks in coastal Oregon, indicating that it occupied the Eastern North Pacific Ocean. The most notable feature of this species are openings on the palate, thought to conduct blood vessels, which lead to baleen in modern mysticetes; this suggests it had both baleen and teeth. Since baleen decomposes and is rarely preserved intact in fossils, the scientists relied on digital reconstructions using CT imaging to search for evidence of baleen in *A. weltoni*. Study of the palate using CT scans by paleontologists revealed separate connections between the blood vessels that would have delivered blood to the upper teeth and those that supplied the baleen. This observation supports the co-occurrence of teeth and baleen, indicating that the tooth-to-baleen transition occurred in a stepwise manner from just teeth, to teeth and baleen, to only baleen.

The toothed baleen whale provides fossil evidence of a major shift in feeding behavior from a single-prey pursuit predator to a bulk filter-feeding mode for obtaining food. Wear facets on the teeth suggest that this whale fed on individual prey items, likely fish. However, the presence of baleen suggests that these whales employed some form of filter feeding. While modern baleen whales are frequently described as feeding on krill, many feed on schools of fish as well, and it is possible that *Aetiocetus* preyed upon single fish and on schools. It is also feasible that whatever "proto-baleen" was present, it may also have simply functioned as an additional structure to secure struggling, slippery fish.

Several features of the skull and jaws suggest that *A. weltoni* was capable of a rapid snapping of the jaws similar to that of some ancient early whales. As judged from the large orbits, vision was a well-developed sense in this species. This species was also well equipped for hearing underwater. Although portions of the skeleton are known, details of the skeletal anatomy have not yet been published.

Studies of other baleen whales from this time show that they were already fully marine—had a tail fluke and flippers. Overall, aetiocetids were small-bodied whales averaging 3–6 feet (1–2 m) in length. A large-bodied species from Japan, twice as big as others, suggests niche partitioning among contemporaneous aetiocetids, and the larger species avoiding competition by pursuing larger prey.

SIZE

Males & females
Length: 10 ft–11 ft
(3 m–3.5 m)
Weight: Unknown

DIET

Fish

HABITAT

Coastal marine

AGE

28–24 mya

An early toothed baleen whale possessed both teeth and baleen. The structure of the proto-baleen is unknown and could have just been exposed bristles rather than organized baleen plates. They likely batch fed on concentrations of krill and fish.

The family Aetiocetidae is the most diverse group of toothed baleen whales represented by six genera and nine species known from both sides of the Pacific during the Oligocene. The evolutionary relationships of aetiocetids indicate that *A. weltoni* is most closely related to *A. polydentatus* from Japan. Aetiocetids lived from 33.9 to 23 mya and are significant early baleen whales preserving an intermediate morphology from raptorial to bulk filter-feeding; species show a diversity of feeding from exclusively raptorial to suction assisted raptorial feeding to raptorial and baleen feeding.

In addition to the Aetiocetidae from North America and Japan, numerous toothed baleen whales have been reported from the Southern Hemisphere (South America, New Zealand, Antarctica, and Australia), ranging from the late Eocene (36.4 mya) to the late Oligocene (23 mya). Toothed baleen whales display a broad variety of feeding strategies although their feeding ecology has been debated (e.g., aetiocetids). The oldest known baleen whale *Mystacodon selenensis* from the late Eocene of Peru is tooth-bearing and has been hypothesized as a suction feeder and possibly benthic feeder with the assistance of the forelimb. Other mostly tooth-bearing baleen whales that show some specialization for suction feeding include *Mammalodon*, *Maiabalaena* (lacking teeth), and possibly the aetiocetid *Fucaia*. The toothed baleen whales, *Janjucetus, Coronodon*, and *Llanocetus*, likely employed the teeth and jaws (also known as raptorial feeding) to obtain prey. As judged by the large number of early baleen whales described in just the last few years, the evolution of feeding strategies in early baleen whales and the transition from teeth to baleen is likely more complex than our current understanding.

# WALKING SEAL

## A recently discovered fossil seal shows that seals and their kin walked on land using webbed feet before they evolved flippers.

It has long been known that pinnipeds (seals, sea lions, and walruses) are closely related to terrestrial carnivores, either bears or weasels, but just how pinnipeds first adapted to a semiaquatic existence has been a mystery. For many years the oldest well-represented fossil pinniped, *Enaliarctos* was a seal-like animal with well-developed flippers.

This mystery is now partially solved. One of the most exciting marine mammal discoveries from Canada's high Arctic sheds new light on the origin of pinnipeds. Dubbed the "walking seal," *Puijila darwini* was found in 24–20 mya lake deposits on Devon Island, Canada. *Puijila* means young sea mammal in the language of the Inuit people inhabiting the fossil's discovery site. The species name *darwini* pays homage to Charles Darwin, who predicted this "missing link" between land to sea animals in his landmark work *On the Origin of Species by Means of Natural Selection* (1859). As Darwin wrote, "A strictly terrestrial animal, by occasionally hunting for food in shallow water, then in streams or lakes, might at last be converted into an animal so thoroughly aquatic as to brave the open ocean."

Evolutionary relationships of the walking seal have been debated. This species was originally proposed to be a transitional pinniped filling the gap between the fully flippered *Enaliarctos* and terrestrial carnivores. Currently, the best supported hypothesis proposes that *Puijila* is either an early pinniped, pinniped relative (Pinnipedimorpha), or an earlier diverging member of a stem group of pinnipeds including *Potamotherium, Kolponomos, Enaliarctos, Pteronarctos* and *Pinnarctidion*. A far-north center of distribution for the walking seal is consistent with the North Pacific occurrences of other marine fossil pinnipeds.

Most of the skeleton of the walking seal is well preserved. The skull is wide and otter-like. The large opening below the eye socket suggests enhanced sensitivity of the snout, useful in both seals and terrestrial (particularly fossorial) carnivorans.

Among fossil pinnipeds where skeletal elements are known, the walking seal is the least specialized for swimming. Unlike pinnipeds, it did not have flippers, and more closely resembled otters in its limb proportions and its possession of a long tail and large, flattened toe bones, which may indicate webbed feet. The presence of enlarged, probably webbed feet, robust forelimbs and an unspecialized tail

SIZE

Males & females
Length: 3 ft 3 in (1 m)
Weight: Unknown

DIET

Fish, rodents, ducks

HABITAT

Fresh water, semiaquatic

AGE

24–20 mya

The extinct walking seal had a long tail and webbed feet and was capable of locomotion both on land and in the water.

suggests that the walking seal swam using all four limbs, its webbed fore and hind feet providing propulsion. It was almost certainly not specialized for swimming underwater as the walking seal used simultaneous pelvic paddling, as seen in the modern river otter. As a quadrupedal swimmer, the walking seal could have given rise to both of the swimming modes of modern pinnipeds, the side-to-side hind limb pelvic paddling of true seals (phocids) and the forelimb paddling seen in fur seals and sea lions (otariids).

Evidence from fossil plants indicates that the discovery site for the walking seal had an environment with a cool, temperate climate around a lake comprising a forest community transitional between a boreal and a conifer hardwood forest where rabbits, rhinoceros, antelopes, and other animals came to drink. Likely lakes would freeze in the winter and the walking seal would have traveled over the land in search of food dispersing to the seas of coastal North America.

# GIANT SPERM WHALE

## Sperm whales in today's oceans lack teeth in the upper jaw and suction feed on squid, but in the past a fossil sperm whale that had huge teeth tore the flesh of its large prey.

The fossil record of sperm whales although rich is mostly represented by isolated teeth and other skull elements. In 2008 an international team of scientists discovered a large skull, 10 feet (3 m) in length, and associated lower jaws and teeth of a fossil sperm whale in middle Miocene sediments in southern Peru. They named the fossil sperm whale *Livyatan melvillei* with reference to the biblical monster, *Livyatan* (Hebrew spelling of *Leviathan*), and after the author Herman Melville and his famous novel *Moby Dick*. The single rooted teeth of *L. melvillei* are more than 1 foot (ca. 30.5 cm) long, 4 inches (10 cm) wide, and are thought to be among the largest of any animal yet known. The teeth were also deeply embedded in the jaw bones for support, and they were interlocking to give the animal an impressive meat-carving, slicing bite that could rip apart large prey, such as baleen whales. This is supported by the discovery at the same site of large numbers of baleen whale skeletons, as well as the remains of other marine mammals, such as beaked whales, dolphins, porpoises, sharks, turtles, and seabirds. Fossils of the giant sperm whale have also been found in Chile indicating that it had a wider distribution in the Southern Hemisphere.

The giant sperm whale represents one of the largest known predators similar to the size of the modern sperm whale. The teeth in both its upper and lower jaws differ considerably from the modern sperm whale, which possesses teeth only in the lower jaw. Rather than catch prey with its teeth near the water's surface, the modern sperm whale uses its powerful tongue to suction feed giant squid at depth. Another sign that this whale was a top predator that had a killer bite is the large opening in the head used to accommodate substantial and powerful jaw-closing muscles. Occlusal wear facets on the teeth support the hypothesis of prey capture by biting and tearing. The snout was short and wide, allowing it to bite more strongly using the front teeth and resist the struggles of its prey.

Rather than being a direct ancestor of the modern sperm whale, scientists think that the giant sperm whale is a close relative. The fossil skull exhibits an enlarged dish-shaped skull like the modern species, which suggests that it may have had a large spermaceti organ, filled with a waxy fluid once sought after by whale hunters, that may have played a role in the production and transmission of echolocation sounds.

SIZE

Males & females
Length: 43 ft–49 ft
(13 m–14.9 m)
Weight: 50 tons
(45 tonnes)

DIET

Large prey including other marine mammals

HABITAT

Coastal marine

AGE

13–12 mya

Unlike the modern squid-eating sperm whale, the giant fossil sperm whale had relatively large teeth in both the upper and lower jaws and may have fed on other sea mammals.

The giant sperm whale was an apex predator and may have competed for food with the giant shark *Carcharocles megalodon*. Although modern whales often hunt cooperatively, given its size *Livyatan* may have hunted alone. Its extinction was probably caused by a cooling event at the end of the Miocene causing changing environmental conditions, including a reduction in almost one half of the diversity of baleen whales, thus impacting the giant sperm whale's prey.

# Walking Sea Cow

## Sea cows today are entirely aquatic and have powerful tails to move in water, but they originated on land and walked using their four legs.

Like whales and pinnipeds, sea cows once walked on land. *Pezosiren portelli* (from the Greek *pezos* for walking and Latin *siren* for mermaid) is a pig-sized animal known from the Eocene of Jamaica. The walking sea cow occupied both land and sea and swam like otters by spinal and hind limb undulations. Like all sea cows, this species was herbivorous, feeding on seagrasses, although it lacked the downturned skull of modern manatees and dugongs. The first upper incisor is enlarged to form a small tusk. *Pezosiren* had a relatively short neck, long barrel-shaped trunk, short legs, and a substantial (but not powerful) tail. The robust limb elements and strong connection between the pelvis and upper leg bone indicate that the walking sea cow could support its body on land, although it shows aquatic specializations (e.g., thick, dense ribs for ballast) that suggest it spent more time in water. As in other sea cows, the ribs are thick and composed of dense bone. The foot bones are short and flattened, resembling terrestrial hoofed mammals rather than the flippers of seals, whales, and modern sea cows. *Pezosiren* is placed in the family Prorastomidae, which is most closely related to *Prorastomus sirenoides* also known from Jamaica. *Prorastomus* and *Pezosiren* were found in coastal and estuarine deposits but other indicators suggest the likelihood of freshwater foraging.

Sea cows are nested within Paeungulata, a group that includes hyraxes, elephants, and their extinct relatives. Many of the earliest fossil records of these groups are found in Africa. Molecular data also support an African root for Paenungulata. Sea cows first appear in the middle Eocene of Jamaica, Florida, and Africa; the oldest record is an ear bone from lake deposits in Africa (Chambi, Tunisia). The Chambi sea cow shows morphological characteristics consistent with an aquatic lifestyle and the capability for underwater hearing. Sea cows, like whales, may have evolved in fresh water; their later dispersal to the North Atlantic (Jamaica) from Africa provides evidence of their subsequent adaptation to marine waters.

SIZE

Males & females
Length: 7 ft (2.1 m)
Weight: 330 lb (150 kg)

DIET

Seagrasses

HABITAT

Semiaquatic

AGE

50 mya

The extinct walking sea cow's distinctive features include a short neck, a long trunk, short legs, and a short tail. It likely spent more time in water than on land.

In addition to prorastomids, a second group of stem sirenians—Protosirenidae—lived during the middle and late Eocene along the Atlantic Ocean, Mediterranean Sea, and Indian Ocean. Protosirenids represent the next evolutionary step and they possessed short forelimbs and reduced hind limbs suggesting that they spent less time than prorastomids on land. Protosirenids begin to show typical sirenian characters, such as a downturned rostrum, indicating that they grazed on seagrasses on the seafloor. Protosirenids show evidence of having well-developed senses of smell and sight in addition to enhanced tactile sensitivity. Some protosirenids exhibit sexual dimorphism.

By the Oligocene both protorastomids and protosirenids were extinct. This epoch was followed by the evolution and diversification of modern sirenians, including manatees (Trichechidae) and dugongs (Dugongidae).

# EXTINCT SEA COW RELATIVE

## A relative of the sea cow, *Behemotops proteus* was named after the biblical monster Behemoth. It fed on marine algae and seagrasses.

*Behemotops* represents the earliest lineage of the only extinct order of marine mammals—desmostylians. They are so-called (from the Greek *stylian* for "pillar") because of their most bizarre feature—bundled columnar teeth resembling a miniature six-pack of cans. Desmostylians had hippo-sized bodies with broad, heavy-hooved feet, and they are found exclusively in marine sediments along the margins of the North Pacific (from Japan to Alaska to Mexico). *Behemotops* is one of seven genera and 12 species grouped into two families: Paleoparadoxidae and Desmostylidae. However, some recent analyses position *Behemotops* as an early diverging form, outside of these families.

*Behemotops* is currently known by two species: *B. proteus* and *B. katsuiei* from the middle or late Oligocene (33.9–23 mya) in North America and Japan. A partly articulated skeleton from Vancouver Island, Canada, is referred to as *B. proteus*. Reevaluation of specimens that originally referred to a third species, *Behemotops emlongi*, were found to differ from *Behemotops* and were renamed *Seuku emlongi*. What is and what isn't *Behemotops* as well as the evolutionary history of desmostylians is still debated. The taxonomy of desmostylians has been hampered by their rarity, the fragmentary remains of noncomparable elements, and variation based on specimens of different growth stages (i.e., adults versus juveniles), necessitating revision of our understanding determined by more complete material.

Compared to other desmostylians, *Behemotops* was relatively small, with *Behemotops katsuiei* having an estimated body length of 9.5 feet (2.90 m), making it the smaller of the two species. *Behemotops* can be characterized as having a long, narrow head, a deeply concave palate, enlarged incisor tusks that point sharply downward, and three small upper incisors on each side with small round roots. Given the close spacing between the roots it seems likely that the crowns of the teeth were narrow. The canine was long, slender, and ventrally curved.

New information about the anatomy of *Behemotops* suggests that rather than a wide, flat-edged snout as originally suggested, it may instead have had a narrow snout. Their low-crowned teeth with rounded cusps are similar to those of early elephants and grazing hoofed mammals, differing from the bundled columnar teeth characteristic of later desmostylians.

SIZE

Males & females
Length: 6 ft 6 in (2 m)
Weight: 220 lb–440 lb (100 kg–200 kg)

DIET

Seagrasses

HABITAT

Coastal marine

AGE

24.8–24.12 mya

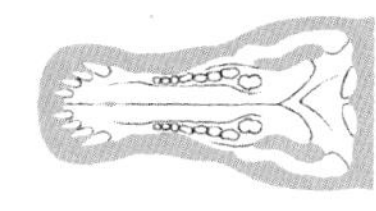

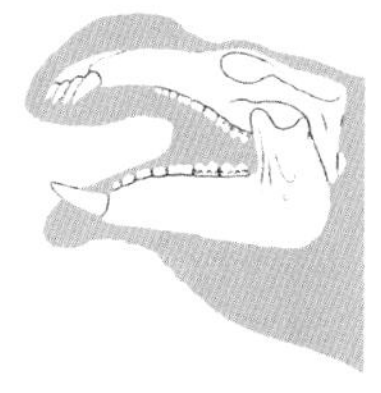

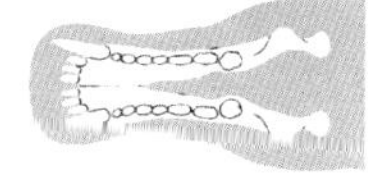

This extinct desmostylian, *Behemotops proteus*, had a long, narrow head and broad, crushing molars used for grazing on seagrasses. They supported their hippo-sized body on heavy limbs as they swam or walked slowly along the seafloor.

A semiaquatic browser, it fed on marine algae and seagrasses. Tooth wear data together with isotopic analysis of the teeth suggests a diet of aquatic plants from shallow coastal bottoms. It is likely that they used their tusks to root up plants and then sucked them in.

Reconstructions of the skeleton and the resulting inferences regarding the locomotion of desmostylians have been both controversial and amusing. Various renditions of these animals have included resemblances to sea lions, frogs, and crocodiles. Desmostylians had an upright posture similar to that seen in some ground sloths and extinct, herbivorous, odd-toed hoofed mammals. The body was probably well off the ground with the limbs under the body. The hind limbs were heavy and it appears that they may have supported their weight on the hindquarters. They were hypothesized as slow-moving, scrambling on uneven and slippery surfaces when moving between land and sea. Locomotion in the water was by forelimb propulsion, resembling polar bears with which they share similar limb proportions. A study of bone histology suggests that *Behemotops*, along with several other genera, exhibited an increase in both bone volume and compaction consistent with shallow-water swimmers that either hovered slowly at a preferred depth or walked on the bottom.

An odd-toed hoofed mammal ancestry for desmostylians from fossils known from the middle Eocene (47.8-38 mya) in India and Pakistan offers a favorable area of origin and dispersal route from Asia to the Pacific Rim. Desmostylians were confined to coastal marine rocks, and ranged as far south as Baja California.

# TOOTHLESS WALRUS

## The toothless walrus, a close fossil relative of the modern walrus, possessed no teeth except for tusks. It likely used them similarly to the modern walrus for social encounters rather than for feeding.

SIZE

Males & females
Length: 7 ft (2.1 m)
Weight: Unknown

DIET

Marine invertebrates; suction feeder

HABITAT

Coastal

AGE

5.3–2.5mya

The extinct, large-bodied toothless walrus (*Valenictus*) is distinguished in lacking all teeth except for the canine teeth that are elongated tusks. Originally described on the basis of an upper arm bone, occurrences of the toothless walrus including skulls and partial skeletons are known from Pliocene coastal deposits of northern and southern California and Baja California. Two species are known: *V. imperialensis* and *V. chulavistensis*, though other records are preliminary and await formal descriptions.

Similar to its close relative the modern walrus, the toothless walrus possessed adaptations for benthic suction feeding, such as an arched palate and reduced dentition that facilitated a molluscan diet. As true for the modern walrus, the tusks of the toothless walrus were likely used for social display and not in feeding.

The lack of post-canine teeth in the toothless walrus parallels the condition in suction feeding toothed whales that have undergone tooth loss, for example, beaked whales, beluga, and Risso's dolphin. The modern walrus does not use teeth to pierce or feed on mollusks; rather it is a specialized suction feeder.

In addition to the lack of cheek teeth, the toothless walrus had a number of skeletal features that distinguished it from other walruses. Unique among pinnipeds are its thick, dense ribs and limb bones that likely provided ballast and allowed the animal to maintain buoyancy in the water. This would have been especially important in occurrences of the toothless walrus in the proto-Gulf of California (Salton Trough) which was hypersaline and made large, heavy animals even more positively buoyant in the water. Without sufficient ballast, feeding on benthic invertebrate in shallow water would have been quite difficult. On the basis of vertebral as well as limb bone anatomy it has been proposed that the toothless walrus was a forelimb/hind limb swimmer like the modern walrus. Morphology of the upper arm bone suggests that it may have relied more on forelimb flexion and pronation during swimming than the modern walrus, though more study is required.

The toothless walrus is most closely related to the modern walrus, *Odobenus*, and is included in the subfamily Odobeninae. During the Pliocene at least four to five walrus species inhabited the temperate-subtropical shorelines and embayments

The extinct toothless walrus (*Valenictus chulavistensis*) was characterized by well-developed tusks likely used in social display rather than feeding, as is true for the modern walrus.

of the eastern North Pacific, which contrast with the single modern walrus species living in the Arctic (see page 108). Loss of walrus diversity coincides with faunal turnover of various marine vertebrates: for example, porpoises replaced by dolphins, archaic baleen whales replaced by modern forms, loss of flightless auks, and giant bony-toothed birds. These faunal changes came about in conjunction with uplift of the California Coast Ranges, loss of shallow embayments, an increase in rocky shore habitat, and rapid sea level changes.

## TUSKS DO NOT A WALRUS MAKE

Since early in their history some walruses lacked tusks. Tusks, which can be either incisor or canine teeth, are defined as ever-growing, projecting teeth made of dentine rather than hard enamel. Tusks are found in a wide array of animals including sea mammals, such as walruses, narwhals, and dugongs, as well as elephants, pigs, warthogs, hippos, and hyraxes. But how did tusks evolve? A recent study revealed that tusks actually evolved in dicynodonts, distant relatives of mammals that lived some 270 mya. As it turns out only mammals evolved true tusks, defined as having surfaces made of dentine rather than hard enamel as in some dicynodonts, and they did so independently many times in the past. In an interesting twist, some elephants are evolving no tusks at all as an evolutionary response to the poaching and killing of them.

# ROBUST-TOOTHED SEAL

## A fossil seal discovered in Peru had large, robust teeth and fed on hard-shelled mollusks rather than fish and squid like modern seals.

SIZE

Males & females
Length: 9 ft (2.7 m)
Weight: Unknown

DIET

Mollusks, echinoderms

HABITAT

Coastal marine

AGE

5.75 mya

A fossil seal, *Hadrokiris martini*, named for its extremely robust skull and teeth (from the Greek *hadros* meaning stout or robust, and the Quechua word *kirus* for tooth) was described from the Sud-Sacaco region of Peru in rocks (Pisco Formation) of late Miocene to early Pliocene age (ca. 10–4.5 mya). The skull, jaws, and teeth are massive. The teeth are the most robust of all living and extinct phocid seals. The robust teeth together with large attachment surfaces for powerful jaw-closing muscles suggests a very different diet than that of most seals who primarily feed on fish and squid. The diet of *Hadrokiris* likely included hard-shelled invertebrates such as bivalve mollusks and echinoderms, especially sea urchins. The extensive degree of tooth wear and fracture supports this interpretation, as does the presence of warm-water crustaceans widely represented in the Pisco Formation. Also, the large neck vertebrae of the robust-toothed seal confirm its powerful neck musculature. This species shows several adaptations (e.g., erect head position) that suggests it had a greater ability to move on land than modern seals, and may have spent more time on shore rather than at sea. Specimens of *Hadrokiris* also exhibit sexual dimorphism.

Two other seal species are known from this locality: *Acrophoca longirostris* and *Piscophoca pacifica*, and a third species, *Homiphoca capensis*, is represented by South African fossils. *Hadrokiris* appears most closely related to *Piscophoca*. All four fossil species are southern seals belonging to the subfamily onachinae also known as monk seals, and they are more closely related to leopard seals than to elephant seals.

The evolutionary history of pinniped communities by paleontologist Ana Valenzuela-Toro and colleagues suggests that the South American record was dominated by phocids (i.e. *Hadrokirus*, *Acrophoca*, and *Piscophoca*) in the past, but today is occupied by otariids (fur seals and sea lions). They suggest that this was due to a rise in sea levels, which would have reduced haul-out sites suitable for phocids, followed by an increase in rocky islands surrounded by a deeper water environment that favors otariid seals.

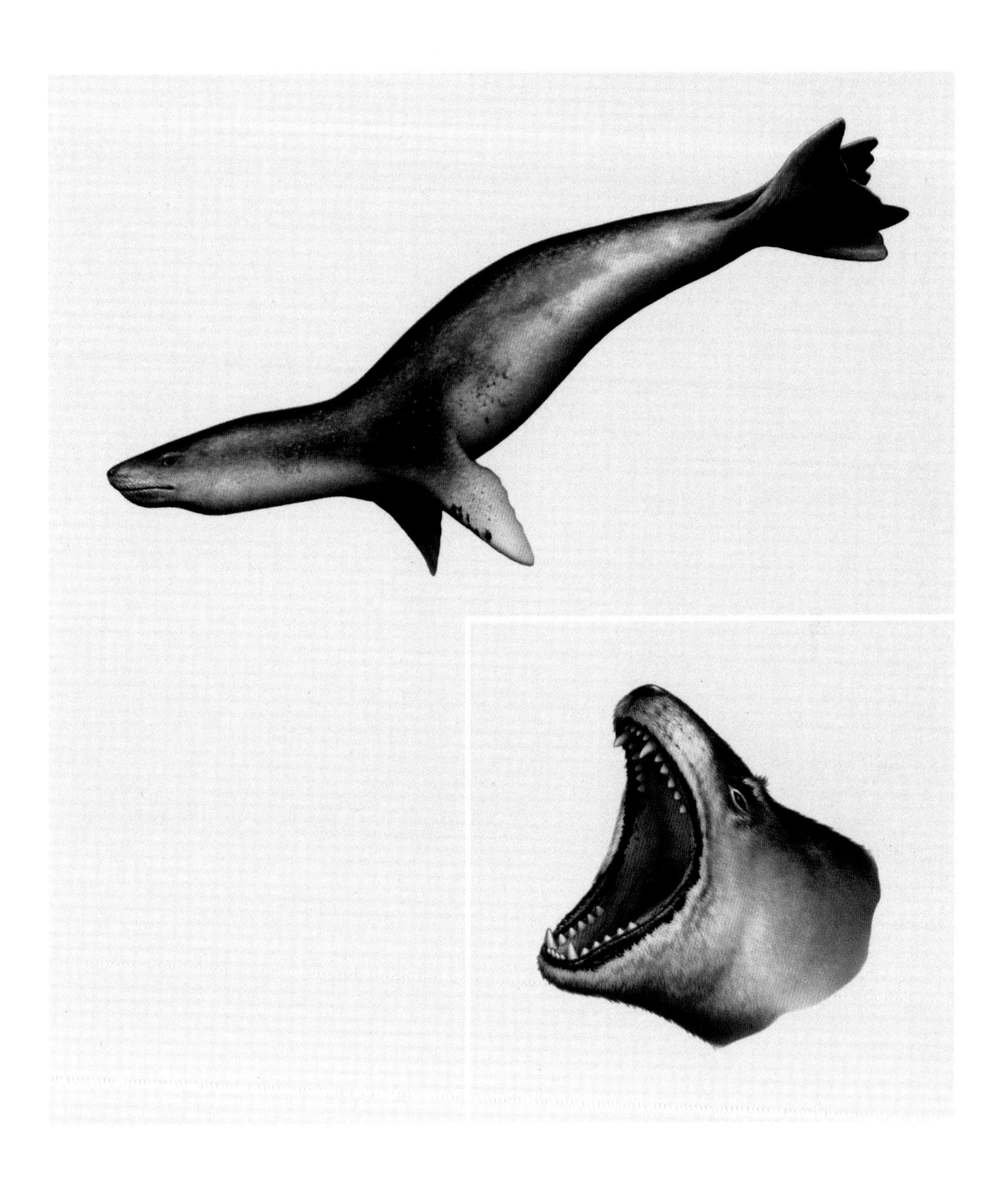

The extinct robust-toothed seal likely fed on hard-shelled invertebrate prey, such as mollusks and sea urchins, using its sharp teeth, rather than fish as is true for modern seals.

# Oyster Bear

## A bizarre fossil marine mammal from the Pacific Northwest is dubbed the "oyster bear" and likely crushed hard-shelled invertebrates with its broad teeth.

The oyster bear (*Kolponomos newportensis*) was originally described in the 1960s from the Olympic Peninsula of Washington by the University of California Museum of Paleontology's Ruben A. Stirton, who initially interpreted this animal as a marine racoon. Later discoveries of complete skulls and jaws and a second species, *Kolponomos clallamensis*, from the early Miocene of Oregon (Nye Mudstone) and Washington (Clallam Formation) was described. Another specimen, an upper jaw fragment from the early Miocene of Unalaska Island, Alaska was referred to cf. *Kolponomos*.

SIZE

Males & females
Length: 4 ft (1.2 m)
Weight: 180 lb (82 kg)

DIET

Mollusks

HABITAT

Coastal marine

AGE

20 mya

The oyster bear had a massive skull with a markedly downturned snout. The eyes were directed anteriorly rather than to the sides of the head, suggesting the oyster bear could see objects directly in front of its head, which would have aided an animal selectively eating rock dwelling benthic or attached invertebrates. The enlarged muzzle, lips, and whisker development suggest that this species had enhanced tactile sensitivity. Large paroccipital and mastoid processes on the back of the skull indicate that the oyster bear had well-developed neck muscles that could have enabled strong downward movements of the head.

One of the most distinguishing features of the oyster bear is the way that it fed. Rather than having teeth that could cut meat like other carnivorans, such as cats, bears, and dogs, it crushed mollusks in its robust jaws studded with broad teeth with rounded cusps. Using multiple lines of evidence, including analysis of bite forces of the lower jaws as well as observations of tooth wear by paleontologists, showed that the oyster bear and the saber-toothed cat (*Smilodon*) independently evolved a similar jaw shape and function. The front of the jaw in the oyster bear is deep and buttressed, giving it a prominent chin like the saber-toothed cat and it may have used a similar strategy of anchoring the head with the lower jaw and then using the leverage to produce a powerful bite similar to the saber-toothed cat but without elongated canine teeth. As described by Riley Black, *Kolponomos* "bit like a saber-toothed cat, crunched like a bear." A unique prey capture-mastication sequence was proposed for the oyster bear that does not have a close analogue in species in modern ecosystems. Initially prey capture involved anchoring and wedging of the

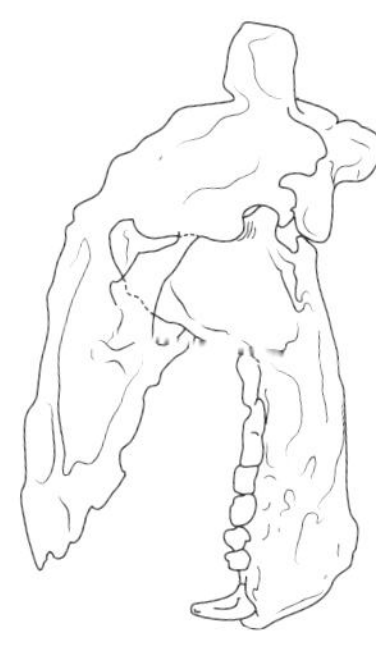

The extinct oyster bear had a massive skull with crushing teeth, strong limb musculature, and robust fingers and toes, suggesting that this animal was adapted for life on land and in the water.

lower incisors and canines between the shelled prey and the substrate, followed by closure of the mouth so that the upper and lower anterior teeth bracketed the shell of the prey. Next, high torque was applied as a fulcrum to dislodge the prey from the substrate, assisted by powerful neck muscles. Finally, in crushing bites, the hard-shelled prey were smashed using their otter-like teeth.

The few known limb and foot elements indicate that the oyster bear was not fully aquatic. Likely it was semiaquatic and capable of movement on land and in the water. The forearm musculature was strong and the robust digits were capable of powerful movement and may have been used to procure food.

The evolutionary relationship of the oyster bear to other carnivores has been debated, allying this animal with pinnipeds or at the base of arctoid carnivorans (bears, racoons, weasels, and related forms). The current best-supported hypothesis places this species as intermediate between bears and bear-dogs, in the Amphicynodontidae, which includes *Amphicynodon, Pachycynodon, Allocyon,* and *Kolponomos. Kolponomos* and *Allocyon* are hypothesized to be the group from which Pinnipedimorpha (seals, sea lions, walruses, and their kin) arose.

# GIANT OTTER

## The giant otter may have evolved in fresh water and followed river systems like modern river otters to disperse to nearshore environments.

Large otters of the genus *Enhydritherium* (from the Greek *enhydris*, "otter" and "beast") are known by two species. *Enhydritherium lluecai* is known from the late Miocene of Europe (Spain). A second species, the giant otter *Enhydritherium terranovae* (from the Latin *terra* meaning "earth" or "land" and *novus* meaning "new"), originally described based mostly on teeth and lower jaw fragments, is now known by additional material including a partial skeleton. The giant otter had a wide geographic range and is reported from marginal marine deposits and freshwater deposits in Florida and California as well as from freshwater deposits in Mexico from the late Miocene to early Pliocene. These discoveries suggest that the giant otter is likely to have spent more time on land than the living sea otter (*Enhydra lutris*) following river systems and traveling overland for considerable distances between water, much like modern river otters *(Lontra canadensis)*.

The skull of the giant otter resembles that of a typical river otter, with a distinctly rounded cranium and relatively broad frontal structure. Features of the nuchal crest and the robust, elongated mastoid, suggest that this species possessed neck muscles more powerful than living otters. The lower jaw is similar to that of *Lutra* and not as massive as in the sea otter *Enhydra*. The thickened tooth cusps of the giant otter and their tendency to show heavy wear suggests that these otters, like the modern sea otter, consumed extremely hard food items such as mollusks in addition to fish.

Although the modern sea otter displays specializations for hind limb propulsion such as elongated distal hind limb elements, the giant otter with its heavily developed upper arm muscle insertion surfaces suggest that unlike *Enhydra* this extinct otter was a forelimb swimmer. The foot bones of *Enhydritherium* are similar to another fossil otter *Enhydriodon* and differ from those of other otters in being short and slender, little longer than the hand bones. In *Enhydra* the foot bones are elongated and flattened as a swimming adaptation. The distal phalanges differ from those of other otters in having extremely large ungual sheaths, suggesting the presence of exceptionally large claws. It is unclear whether *Enhydritherium* like *Enhydra* possessed retractile claws. With more balanced forelimb/hind limb proportions *Enhydritherium* was likely more efficient at locomotion on land.

SIZE

Males & females
Length: 3 ft–4 ft
(1.2 m–1.5 m)
Weight: 50 lb–100 lb
(22 kg–45 kg)

DIET

Fish and shellfish

HABITAT

Nearshore marine; estuarine and inland fresh water

AGE

6.5–4.5 mya

The extinct giant otter swam by using its forelimbs rather than by the hind limb propulsion that is seen in modern otters.

The giant otter is thought to be related to both the Old World otter *Enhydriodon* and the modern New World sea otter *Enhydra*—all are assigned to the river otter subfamily Lutrinae in the family Mustelidae, including skunks, ferrets, martens, and minks.

# ECHO-HUNTING WHALE

## Discovery of the echo-hunting whale, found in 25-million-year-old deposits in South Carolina, US, suggests that echolocation evolved early in whales.

*Echovenator sandersi* is the earliest whale to have evolved echolocation, the ability to produce high-frequency sounds that bounced off prey, creating echoes received by the inner ear. The name *Echovenator* comes from the Latin for "echo hunter" referring to echolocation. This small toothed whale is a member of the family Xenorophidae, the most basal group of toothed whales and a distant relative of modern dolphins. *Echovenator* is represented by a nearly complete skull that was found in nearshore marine deposits of the Oligocene Chandler Bridge Formation in South Carolina. Paleontologists analyzed high-resolution CT scans of *Echovenator*'s well-preserved ear and uncovered many features, such as a loosely coiled cochlea, a spiral cavity of the inner ear that contains hearing receptors; this is found also in today's dolphins that can hear high-frequency sounds. Other features of the skull of *Echovenator*, such as bony correlates for facial air sacs and expanded attachment areas for facial musculature, support its capability for echolocation although in a less specialized form than modern toothed whales. Study of the facial region of the skull in another early toothed whale fossil *Cotylocara macei* demonstrate that xenorophids already had the ability to produce echolocation sounds. These early toothed whale fossils confirm that the Oligocene (30–25 mya) represents an important period of whale evolution with fossils documenting the initial diversification of both toothed and baleen whales from stem whales, ancestral Eocene archaeocetes.

Echolocation was a key innovation, which is one of the major reasons why toothed whales were so successful, so its early evolution is significant. As determined from studying the ears of *Echovenator*, it appears that the ability to hear high-frequency sounds evolved before the ability to produce high-frequency sounds, and echolocation became even more specialized in toothed whales. Baleen whales that do not echolocate and are specialized to produce and hear low-frequency (infrasonic) sounds used in long distance communication lost some of their initial specializations for hearing high-frequency sounds. These results are contrary to the traditional hypothesis that toothed whales evolved from low-frequency hearing specialists. Whales started becoming sensitive to high-frequency sounds before the two major groups alive today—the toothed whales and baleen whales—split from each other.

SIZE

Males & females
Length: 4 ft 10 in
(1.48 m)
Weight: Unknown

DIET

Fish

HABITAT

Coastal marine

AGE

c. 27–24 mya

The extinct echo-hunting whale had well-developed hearing receptors in its inner ear. It is believed to be the earliest whale to have evolved echolocation.

The evolution of toothed whales is closely tied to the origin of echolocation. Scientists have suggested that echolocation in early toothed whales was initially an adaptation for feeding at night on vertically migrating mollusks, especially nautiloids (shelled cephalopods, relatives of squid and octopuses). Echolocation appears to have coevolved in predator (toothed whales) and prey (nautiloids). Evidence for this is the production of stronger echoes by gas-filled nautiloids compared to soft-bodied cephalopods, such as squid. These nautiloids would have been easily detectable prey for early echolocating toothed whales and may be responsible for their near demise 25 mya. Subsequent modification of echolocation with finer resolution, as documented in the fossil record, was driven by whales hunting squid and other prey in deep water.

## Echolocation: Seeing in the Dark with Sound

Unlike baleen whales, toothed whales locate and hunt in dark and murky environments, at night or at great depth, using echolocation. The ability to navigate using high-frequency sounds well above the range of human hearing, and reception of its echo, is called echolocation. Toothed whales are rivaled only by some bats in their capacity to produce and hear ultrasonic sound, typically between 120–180 kHz. For comparison, humans hear in the range of .02–17 kHz. Unlike humans, where sounds are produced in the larynx or voicebox, the echolocation sounds of toothed whales are produced in the nasal region. The sounds are emitted through the "phonic lips"—paired, fat-filled structures, suspended within muscles and air spaces near the blowhole. The phonic lips are embedded in a pair of anterior and posterior bursae that help direct the sound beam. The melon acts as a lens to focus sounds. The returning echoes pass through and under the lower jaw fat bodies before being transmitted to the middle and inner ear. This remarkable adaptation is so precise that toothed whales can distinguish very small differences in the size and type of prey they are targeting, and can also penetrate into sand or mud to locate buried prey.

Toothed whales feed mostly on fish and squid; however, some, such as killer and false killer whales, hunt and eat larger prey such as other marine mammals (see page 92).

The independent evolution of echolocation in whales and bats are excellent examples of convergent evolution where organisms that are not closely related resemble one another. This adaptation allowed both groups to take advantage of unexploited food resources. Although different in details, the echolocation abilities of bats and toothed whales rely on the same changes at the molecular level—the presence of the gene prestin—a remarkable example of independent acquisition in two diverse groups. The discovery of prestin in echolocating species and its function as an amplifier for high-frequency sounds is significant, since there is concern about how noise pollution affects hearing in whales. Exposure to high-intensity sound for long periods of time can damage hearing neurons, affecting echolocation, which toothed whales rely on for orientation and feeding. Ultimately, this may lead to hearing loss, which has been found to result in stranding and death in some whales.

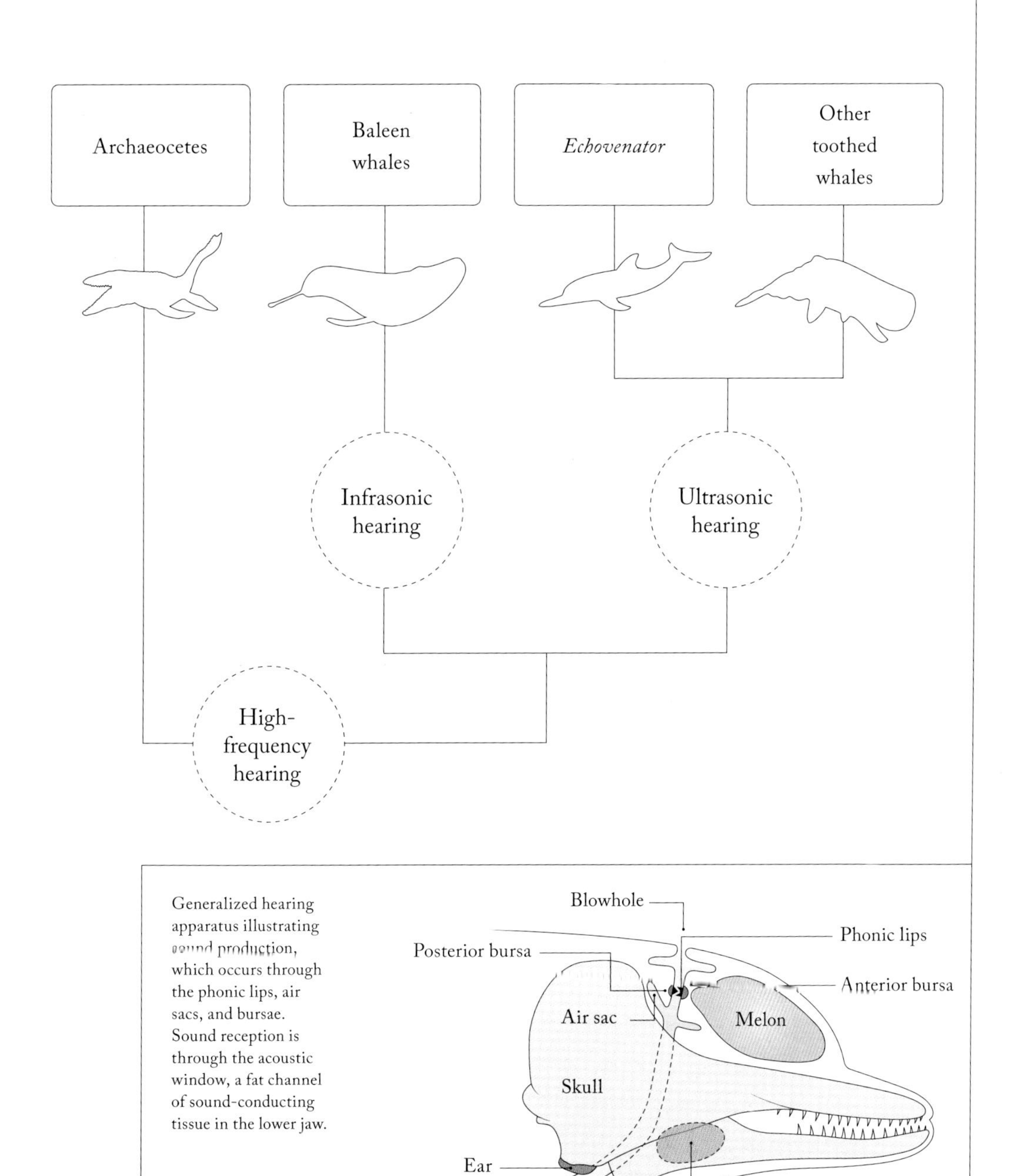


Generalized hearing apparatus illustrating sound production, which occurs through the phonic lips, air sacs, and bursae. Sound reception is through the acoustic window, a fat channel of sound-conducting tissue in the lower jaw.

# 2
# DISCOVERY

Whether reporting a new animal previously unknown to science, or collecting fossils using nontraditional means, significant discoveries of extinct and living sea mammals have been and continue to be made around the world.

Among the species selected for this chapter is an early whale from the southeastern US whose large bones were originally thought to be from a dinosaur. In a highly unusual method of discovery, fossil dolphins and beaked whales not previously known were dredged from the seafloor by fishing vessels in the North Sea and the Atlantic. Another remarkable discovery is a fossil whale skull embedded in limestone from Egypt that was being cut into decorative slabs at a commercial quarry. Also reported is a fossil whale graveyard in Chile discovered during the expansion of the Pan-American Highway. With no time to remove specimens, it was critically important to record a snapshot of the discovery site using laser imagery that enabled later study of specimens. Amazingly, the largest animals on Earth continue to be regularly discovered, including several new species of both toothed and baleen whales from the Gulf of Mexico and the North Pacific.

A novel yet controversial hypothesis proposed that the mystery of the origin of the modern pygmy right whale was solved and that it was a member of a fossil group long thought to be extinct.

PAGE 48
An African manatee cub and its mother (foreground) (*Trichechus sengalensis*), seen at Chimelong Ocean Kingdom in Zhuhai, China. Like all marine mammals that live in the water, the manatee must surface to breathe.

Mermaids have long been associated with sea cows and this legend when traced back was found to be based on early reported observations of both manatees and dugongs. Sea unicorns, another legend, most likely refer to the narwhal's uniquely developed spiral tusk.

A new fossil walrus discovery is based on a nearly complete skeleton from southern California, adding to the past diversity of more than 20 walrus species in contrast to the present-day single Arctic species. The report of an extinct seal relative from California adds new information to our knowledge of the decline and extinction of a fossil lineage, suggesting that declining temperatures played a role as did competition with rapidly diversifying walruses. Discovery of the coexistence of several sea cow species at different times and places in various parts of the world suggest that they avoided competition by dividing up limited resources, analogous to ecological interactions among coexisting species in modern communities.

# "King of Lizards"

## Originally identified as a dinosaur and described as the "king of lizards," this species is not a reptile but a whale.

SIZE

Males & females
Length: 60 ft–70 ft (17 m–20 m)
Weight: Unknown

DIET

Large fish

HABITAT

Nearshore marine

AGE

35–34 mya

Since Aristotle's time (384–322 BCE) whales have been recognized as mammals and not fish because they have hair, lungs, lack gills, suckle their young by means of mammae, and give birth to live young. During the Renaissance, the Italian artist and engineer Leonardo da Vinci is credited with a description of the first whale fossil found in a cave in Tuscany, Italy, in the fifteenth century. But it was not until the eighteenth century that a fossil whale was first named, although at the time it was thought to be a gigantic lizard.

A skeleton of what was believed to be a large sea serpent was found in Alabama in 1842; it had a partial skull, forelimb, and vertebral column extending nearly 65 feet (20 m). This discovery prompted the German fossil collector Albert Koch to visit Alabama and purchase the bones from farmers. He strung the bones together and reconstructed a single specimen nearly 114 feet (35 m) long; it had a long body, giant flippers, and a huge mouth. He claimed that such a skeleton belonged to a "sea serpent" that he named *Hydrarchos sillimani* after Yale professor and friend of Koch, Benjamin Silliman. The "skeleton" was exhibited in New York and Europe and thousands flocked to see it; most thought that it proved the existence of sea serpents. Later, more careful examination showed that the skeleton in question was a composite of at least five different specimens. Other similar bones including a large vertebra or backbone from the Eocene of Louisiana were published in 1834 by Richard Harlan, American geologist and paleontologist. However, he also did not recognize his find to be from a whale. Rather, he identified it as a dinosaur and proposed to call the animal "the king of lizards" or *Basilosaurus*.

Among additional fossil remains of similar animals found on an Alabama plantation were teeth that were examined by the well-known English anatomist Richard Owen. In 1839, Owen proposed a new name for the teeth and bones, which he recognized as deriving not from a dinosaur or other reptiles but from an extinct whale which he named *Zeuglodon cetoides*, choosing the genus name in recognition of the yoke-like shape of its back teeth (*zeugleh* means "yoke" in Greek and the Latin *dens* means "teeth"), and the species name in recognition of the appearance of its whale-like vertebrae (*cetoides*). However, the first published name takes precedence, in this case, rendering these specimens *Basilosaurus cetoides*.

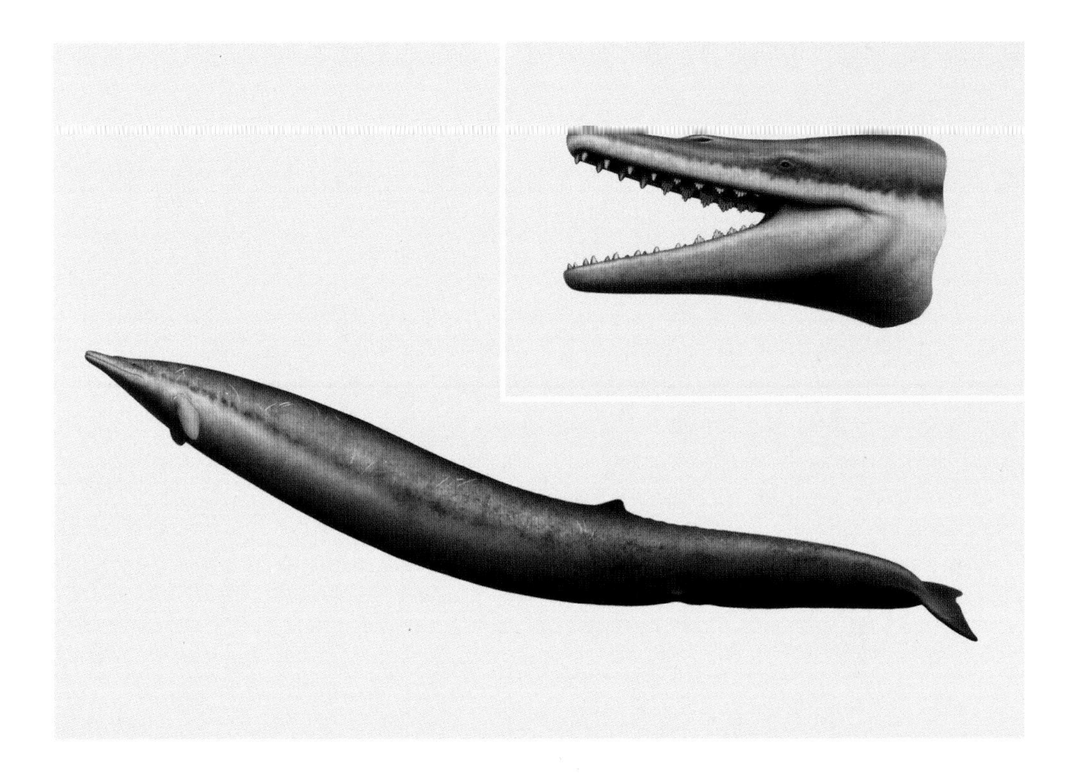

The extinct large stem whale (*Basilosaurus isis*) was the top marine predator of its day, more than 35 million years ago. It had large teeth and a powerful bite, and likely fed on the bones of other cetaceans.

Basilosaurids are recognized today as a family of stem whales comprised of approximately 13 genera and 19 species. They grew to 60–70 feet (18–20 m) and possessed greatly reduced hind limbs with well-formed legs and feet. They were one of the earliest fully aquatic cetaceans. Study of tooth wear in basilosaurids suggests they were large predators with a forceful, crushing bite feeding on the bones of other cetaceans. The genus *Basilosaurus* includes three species that lived from 45 to 35 mya; fossils were found in rocks along the shores of the Tethys Sea (North Atlantic coast of southeastern US), West Africa (Egypt), and possibly Pakistan. *Basilosaurus cetoides* is known from Georgia, Alabama, and Mississippi and it is recognized as the official state fossil of Alabama and Mississippi. The slightly smaller *Basilosaurus isis* is known from many skeletons in Egypt, the majority collected from Wadi Al-Hitan "Valley of Whales" in the Fayum desert, now an internationally known and much-visited UNESCO World Heritage Site (see page 16). Basilosaurids lived in warm, tropical seas of the Eocene, becoming extinct at the transition between the end of the Eocene and the beginning of the Oligocene, a time of climatic change when temperatures cooled.

# Pygmy Right Whale

## The pygmy right whale is the most cryptic and least known in terms of morphology and behavior of the living baleen whales.

*Caperea marginata* is commonly called the pygmy right whale. The name *Caperea* means "wrinkle," referring to the wrinkled appearance of the ear bone, while *marginata* refers to the dark border around the baleen plates of some individuals. It is the sole member of the family Neobalaenidae. Its name pygmy right whale is a misnomer since it differs significantly from right whales (family Balaenidae). It is the smallest baleen whale and has a distinctive skull and skeleton that features large horizontal processes on the vertebrae, and overlapping ribs that restrict mobility in the chest region. It is also the only living whale lineage confined to the Southern Hemisphere, where it is rarely seen. We know more about the pygmy right whale from dead individuals than from live ones.

Pygmy right whales are slender, resembling the streamlined balaenopterids rather than the chunky right whales and bowhead. They have dark gray backs that shade to white on the belly and a pair of chevron-shaped lighter patches behind the eyes. Two grooves are located on the throat similar to the throat grooves of the gray whale. Although the rostrum is arched it is not as conspicuous as in balaenids. They have a small curved dorsal fin located near the posterior end of the body. Like other baleen whales, the ear of *Caperea* was specialized for hearing low-frequency sounds; however, its hearing limit is relatively high.

Although it is not known whether this species is migratory it has a circumpolar distribution in both coastal and oceanic waters within temperate regions. When seen they are usually located within sheltered bays. The whale is typically sighted alone or in pairs. They do not exhibit behaviors commonly seen in other whales such as spy hopping or breaching. Although there is little information on diet, pygmy right whales are skim feeders, and they are known to feed on copepods and krill.

The evolutionary relationships of the pygmy right whale have long been debated. *Caperea* has been allied either with rorquals and gray whales (families Balaenopteridae and Eschrichtiidae) using molecular data and right whales (Balaenidae) based on morphology. Molecular analysis, combined with studies of their bone structure (especially ears) from fossils and other remains, has revealed a third hypothesis, that it is the only survivor of a lineage (family Cetotheriidae) thought to have gone extinct 2 mya. Because of this, researchers refer to the

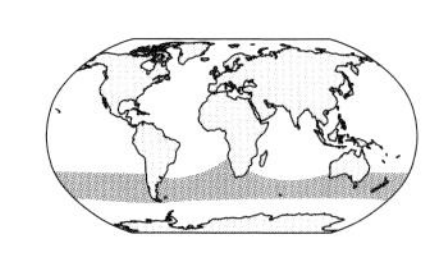

SIZE

Males & females
Length: 20 ft–21 ft
(6 m–6.5 m)
Weight:
7.25 tons–8.5 tons
(6.3 tonnes–7.26 tonnes)
Females longer than males at birth by 6 ft 6 in (2 m)

DIET

Copepods and euphausiids

HABITAT

Coastal and oceanic

LIFE HISTORY

Sexual maturity may occur at lengths greater than 16 ft 4 in (5 m)
Gestation:
10–12 months
Reproduce: 2 years

STATUS

Least concern

The pygmy right whale is the smallest and most elusive of the baleen whales. It has a narrow, downturned rostrum and a pale light-colored chevron near the level of the flippers.

pygmy right whale as a "living fossil," a living species that resembles an ancestral species known only from the fossil record. According to the cetotheriid hypothesis *Caperea* reportedly shares several ear characters with some cetotheres. However, disagreement about whether these shared features represent common ancestry, convergence (independent acquisition), or primitive similarities is at the crux of the matter and will require further analyses of characters and fossils.

Prior to 2012 fossils of the *Caperea* lineage were unknown. Since then, new specimens of this lineage have been found in the late Miocene (up to 10 mya) rocks of Australia, Argentina, and Peru. Recent description of two Pleistocene *Caperea*-like fossils in the Northern Hemisphere from Italy (1.8 mya) and Japan (less than 1 mya) provided the first evidence that *Caperea* crossed the Equator, probably during the brief interglacial intervals associated with pronounced cooling that accompanied Northern Hemisphere glaciation. With falling temperatures, waters near the Equator cooled and became richer in nutrients, which would have made it easier for *Caperea* to extend its range beyond the tropics into the Northern Hemisphere. Other marine mammals including some dolphins, right whales, and elephant seals have made similar journeys. However, as the glacial period was followed by an interglacial period, tropical seas warmed and productivity declined. What was once a tropical gateway became an impassable barrier, leaving populations trapped on either side of the Equator in a warming ocean. Some divided marine mammal populations evolved into separate species such as right whales and elephant seals, but others declined to extinction in one hemisphere, like *Caperea* in the north.

The pygmy right whale was never exploited during commercial whaling, probably because of its small size.

# Early Whale

## One of the most unusual fossil whale finds is a cross section of a skull discovered by a quarry worker in a limestone block.

In 2003 the owner of a marble-cutting company in Tuscany, Italy, had just acquired a block of Egyptian marbleized limestone. After slicing the block into slabs, he discovered the bones of an animal. Thinking that he had discovered a dinosaur, he contacted paleontologist Giovanni Bianucci at nearby University of Pisa. Bianucci recognized the bones as belonging to an early whale. However, before he could provide a more specific identification he needed to know more about where the limestone slabs had come from. The slabs were marketed as "Sheikh Fadl limestone." With knowledge that Sheikh Fadl was a city in Egypt, Bianucci discovered several limestone quarries in the Tarfa Valley east of the city. Bianucci and collaborator Philip Gingerich from the University of Michigan traveled to the limestone quarry and discovered that the rocks were 40 mya. Bianucci and Gingerich named the fossil whale that nearly wound up as a countertop *Aegyptocetus tarfa*—"Egyptian whale from Tarfa." Once the slabs containing bones were put back together, Bianucci and Gingerich were able to reconstruct the skull and the associated partial skeleton. The new fossil species was identified as a protocetid, a group of early whales. The first of this group to be found—*Protocetus atavus*—was also located at an Egyptian rock quarry by workmen who were extracting building stone.

Although the limbs are missing, further study of the exceptionally preserved skull of *Aegyptocetus* revealed that it had well-developed turbinal bones, providing evidence for a highly developed sense of smell, a feature not seen in modern whales. Both the skull and jaws show adaptations for hearing in water, such as a downward slope to the skull, thin-walled ear bones, and a large canal in the lower jaw that helps channel sound to the ears. Long neural spines on the anterior thoracic vertebrae indicate that *Aegyptocetus* was able to support its weight on land like other protocetids. The combination of terrestrial and aquatic characters

SIZE
Males & females
Length: Unknown
Weight: 1,433 lb
(650 kg)

DIET

Fish

HABITAT

Coastal

AGE

41–40 mya

An extinct protocetid whale had powerful jaws and teeth. The well-developed hind limbs and tail were used in locomotion, which suggested that it was active on both land and in the water.

indicate that *Aegyptocetus* was probably semiaquatic. The bones also suggest how it died. Four of the ribs contain tooth marks of a shark, which must have swum upward at the animal and bitten it on the left flank. Today's white shark uses the same strategy and other fossil whales bear the marks of similar attacks by predators.

# NARWHAL

## Known as "sea unicorns," narwhals are toothed whales that have uniquely evolved a long spiral tusk protruding forward from their upper lip.

Sea unicorns with their prominent single horn are mythological animals, typically pure white horses that have a spiral tusk; they can be traced back to classical times. Unicorn horn was traded through the Middle Ages and Renaissance by Vikings. The unicorn horn had magical powers ascribed to it and it was believed to ward off disease, so parts of the tusk were worn as necklaces or ground up and prescribed as medicine. In one of the most ostentatious displays of power and wealth, the Danish Royal Throne chair made in the mid-seventeenth century and housed today in Copenhagen is constructed of numerous unicorn "horns."

The source of the horn is actually the elongated, up to 9 feet (2.7 m) long, spiral tusk (canine tooth) that protrudes from the left upper lip of a male narwhal, though some males develop two tusks. Females sometimes develop small tusks, too, but they usually remain embedded in the upper jaw. It is likely that the unicorn was based on findings of the tusk of the modern narwhal (*Monodon monoceros*), meaning one tooth, one horn, hence the name "the unicorns of the sea." Narwhals are restricted to the Arctic, occurring mostly above the Arctic Circle. They inhabit the Atlantic portion of the Arctic and have a few records of discovery in the Pacific portion.

Various functions have been proposed for the male narwhal tusk, ranging from its use in hunting, or as an environmental sensor or, as most recently proposed, a sexually selected trait developed in males to attract females, similar to a peacock's tail or deer antlers. It is also possible that the tusk has several functions. The environmental sensor hypothesis is based on study of the structure of the tusk, which contains nerve endings of the inner core that connect to the outer surface and makes the tusk sensitive to temperature and chemical changes in the water, using them to find food and helping them to locate females ready to mate. Another possibility is that sensing the environment might help the whales survive the harsh and ever-changing conditions in the Arctic. Video footage has also shown that the tusk is used in hunting prey; narwhals are seen striking Arctic cod with their tusks to stun them before eating.

SIZE

Males: length up to 16 ft (4.9 m); weight up to 1.7 tons (1.54 tonnes)
Females: length up to 14 ft (4.3 m); weight up to 1.1 tons (1 tonne)
At birth: length 5 ft (1.5 mm)

DIET

Fish, squid, and shrimp

HABITAT

Coastal

LIFE HISTORY

Sexual maturity
Males: 12–20 years
Females: 8–9 years
Gestation: 13–14 months
Reproduce: 3 years

STATUS

Least concern
Threat from hunting

ABOVE
A pod of male narwhals, located at Baffin Island, Canada, shows the distinctive black spotting of the body especially around the head, sides, and back.

RIGHT
A narwhal raises its distinctive tusk out of the water.

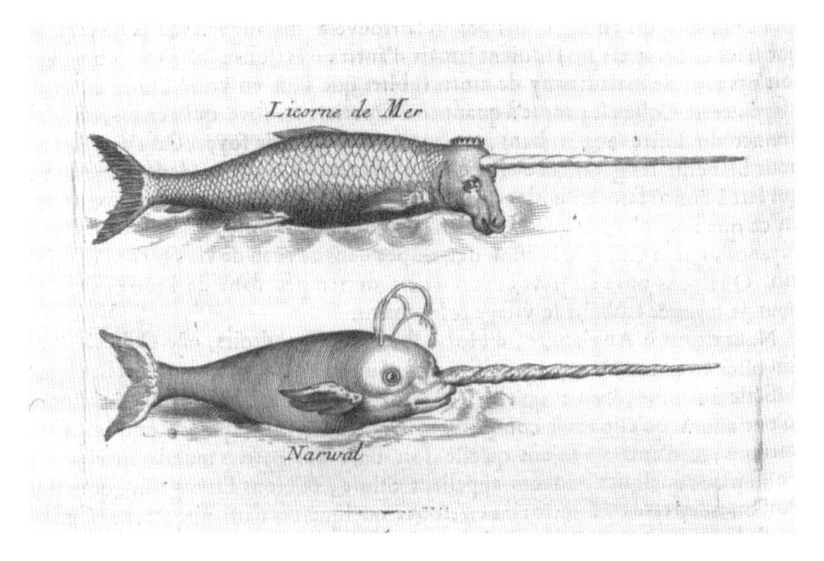


LEFT
Unicorns (top) and narwhals (bottom) have long been linked. During medieval times narwhals were hunted and their tusks sold as "unicorn horns."

OPPOSITE
Narwhals often raise their tusks in order to "cross swords," sparring with other males for access to females, located at Baffin Island, Canada.

Apart from the distinctive tusk in males, narwhals are characterized by a rotund body, small bulbous head, short blunt flippers, and notched flukes. Narwhals are born gray to brownish gray and they darken to all black with white mottling as they age. Older animals appear nearly white with some black mottling near the front and top of head and appendages.

Narwhals are an Arctic species and are found mainly above the Arctic Circle. They mostly occupy the Atlantic portion of the Arctic although they are occasionally recorded from the Pacific. They migrate annually from open waters in the fall to inshore waters in the spring. In summer narwhals follow the ice along the coast and in winter they remain in pack ice.

Most pods of narwhals consist of two to ten individuals. But there is some evidence that there are groups of larger herds of hundreds or thousands of individuals. There is some age and sex segregation, including all male groups, and nursery groups form commonly.

The diet of narwhals is mainly fish, squid, and shrimp. They mostly feed in the winter in deep water. Dives of 5,000–6,600 feet (1,500–2,000 m) and up to 25 minutes in duration have been reported.

Hunting is the major threat to these animals and native hunters in Canada and Greenland hunt narwhals for their tusks and skin. Narwhals are also victims of ice entrapment, which blocks their escape to the open ocean, and happens more frequently today due to climate warming.

OVERLEAF
Narwhal pod located at the Arctic's ice edge.

# Dugong and Manatees

## Mermaids are based on legends in many cultures of sirens, mythical women that had fish-like bodies resembling a group of sea mammals—manatees and dugongs.

Mermaids are represented as contradictory beings, sometimes even within the same culture. In folklore they are often depicted as dangerous creatures, associated with floods, storms, shipwrecks, and drownings. They have also been portrayed as beautiful creatures, who through the melody of their songs lure sailors into the sea and devour them.

Variations of the word "siren" means mermaid in many languages, and it is the source of the larger group to which manatees (family Trichechidae) and dugongs (family Dugongidae) belong in the order Sirenia.

Since classical antiquity to modern times historical sightings by sailors may have been the result of misunderstood encounters with these aquatic animals. When Christopher Columbus set out to sea in 1492 he had a mermaid sighting of his own, and this encounter was actually the first written record of manatees in North America. Off the coast of Haiti he caught a glimpse of "mermaids," writing in his journal, "On the previous day [8 Jan 1493], when the Admiral went to the Rio del Oro [Haiti], he said he quite distinctly saw three mermaids, which rose well out of the sea...." Indeed, manatees and dugongs are both known to rise out of the sea like the alluring sirens of Greek myth, occasionally performing "tail stands" in shallow water. Likely Columbus saw the Antillean manatee (*Trichechus manatus manatus*), one of two subspecies which ranges from northern Mexico to northern South America including the Caribbean islands.

A Portuguese Renaissance historian António Galvão disseminated observations made in the Spanish New World and wrote referring to the Antillean manatee in 1497, "Is there a fish called manatim; is big and has a cow's head and face, and also in the flesh it looks very like it (...) and the female has breasts with nipples that feeds its children who are born alive." In another example, sixteenth-century missionary and naturalist José de Acosta (1590) refers to the Antillean manatee, "In the islands of Barlavento, namely Spanish Cuba, Puerto Rico, Jamaica, there is the so-called manatee, a strange kind of fish, if one can name fish to an animal, whose cubs are born alive, and has teats, and with milk they are raised, and eats herb in the fields; but indeed, usually resides in the water." European descriptions of the Antillean manatee are based on the botanist Carolus Clusius' knowledge of the Caribbean and his publication illustrating the animal in 1605.

SIZE

Manatee: length 8 ft–11 ft (2.5 m–3.5 m); weight 0.5 ton–1.5 tons (0.45 tonne–1.4 tonnes)
Dugong: length 10 ft (3 m); weight 0.6 ton (0.54 tonne)

DIET

Seagrasses

HABITAT

Shallow subtropical-tropical coastal waters; brackish, marine, and fresh water

AGE

50 mya–recent

The West African manatee (*Trichechus senegalensis*) is a vulnerable species, which ranges from Senegal to Angola. It feeds on aquatic plants using its flexible snout and mobile lips to bring food into the mouth.

LEFT
A nineteenth-century hand-colored engraving of a dugong from the Scottish naturalist William Jardine's *Naturalist's Library*. Such illustrations indicate that naturalists were intrigued by these unusual looking creatures.

BELOW
Aboriginal rock art depicting a dugong. This extraordinary artwork was found at Bathurst Head, Queensland, Australia, suggesting that dugongs were common in this area for very many years.

A second subspecies, the Florida manatee (*Trichechus manatus latirostris*) occurs in waters of the southeastern US. Following Columbus's expedition to the Americas, sideshows in Europe advertised "recently discovered" mermaids from the new world, often a dead sirenian.

A description of a mermaid in England's *Magazine of Natural History* is as follows: "A short time back, the skeleton of a mermaid, as it was called, was brought to Portsmouth, which had been shot in the vicinity of the island of Mombass (Kenya). This was allowed to be submitted to the members of the Philosophical Society, when it proved to be the Dugong … It was, if I recollect right, about six feet long: the lower dorsal vertebrae, with the broad caudal extremity, suggested the idea of a powerful fish-like termination; whilst the fore legs, from the scapula to the extremities of the phalanges, presented to the unskillful eye an exact resemblance to the bones of a small female arm." A species of manatee does occur in Africa, *Trichechus senegalensis*, but on the Atlantic side ranging from the Senegal/Mauritania border to central Angola.

In the seventeenth century the Italian missionary Giovanni Cavazzi gives a good description of the so-called fish woman inspired by observations of the African manatee: "There is one [fish] that Europeans call fish-woman and local name is Ngulu-maza [literally, Kikongo or pig water], beautiful name, but so horrendous. Has the muzzle gaping but small in comparison with another that appears to be a male. I think this is the famous triton from fables of mythology, the female may be considered the naiad of the old."

A third manatee species occurs in South America, the Amazonian manatee (*Trichechus inunguis*), an exclusively freshwater species found in the waters of the Amazon River and its tributaries and lakes in Brazil, Guyana, Columbia, Peru, and Ecuador (see page 174). Likely the earliest illustration of this animal was by Brazilian missionary and Frei Cristóvão de Lisboa in 1647, who presented a brief description of the Brazilian manatee accompanied by a good illustration.

Thousands of miles from the seas Columbus sailed, the dugong—found in the Pacific and Indian Oceans—had been living in legend for centuries. In 1959, 3,000-year-old cave drawings were discovered depicting dugongs—the word translates to "lady of the sea" in the Malay language—inside Malaysia's Gui Tambun Cave. In Palau, a Pacific nation of 340 islands, the dugong plays a central role in traditional ceremony and lore. Stories of young women transformed into dugongs endure, and wooden storyboard carvings illustrate dugongs aiding fishermen lost at sea.

# SEA COW COMMUNITIES

## Communities of multiple species of sea cows lived together in various ocean regions during different times in the past by partitioning the resources.

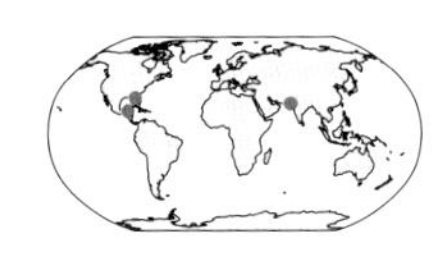

SIZE

Males & females
Length: 5 ft–13 ft
(1.5 m–4 m)
Weight: 615 lb–4,103 lb
(279 kg–1,866 kg)

DIET

Seagrasses

HABITAT

Tropical and subtropical shallow waters

AGE

26–3 mya

Sirenians or sea cows are a group of marine mammals that includes manatees (three species) and the dugong (two species). The world's oceans are characterized today by a single sea cow species living separately in various regions—the dugong occupying the Indian Ocean and manatee species living in the Atlantic. However, in the past it was more common to find several species of sea cows living together. This discovery, termed "multispecies assemblages" by paleontologists Jorge Vélez Juarbe, Nick Pyenson, and Daryl Domning, spurred research to determine the composition, anatomy, and behavior of sea cows in these assemblages.

Vélez-Juarbe and colleagues examined three localities, each from separate time periods (from the late Oligocene of Florida, the early Miocene of India, and the early Pliocene of Mexico) where more than one sea cow species coexisted. They then determined how the species interacted with one another in their environments. To do this, they examined anatomical features that were proxies of dietary preferences in each species. For example, the degree of deflection of the rostrum provides information on whether they were feeding throughout the water column (less downturned rostrum) or were obligatory bottom feeders (more downturned rostrum). Another feature examined was tusk anatomy (size and shape), which provided information on the maximum rhizome or plant stem size that could have been uprooted. Some of these characteristics together with body size estimates (which revealed the amount of seagrasses that could have been ingested) were used to determine how these multispecies assemblages partitioned the resources. Their findings imply that seagrass beds, the animals' principal food source, were different than they are today. Next, they analyzed the evolutionary relationships of all the studied sea cows. The resultant tree showed that the anatomical characters evolved independently in the different sea cow species assemblages. Each fossil assemblage showed species that differed in at least one anatomical feature—rostral deflection, tusk size, and body size allowing their coexistence.

Results indicated that the Oligocene dugongid species from Florida differed in body size and tusk anatomy. The two taxa that have the most similar body sizes (*Crenatosiren olseni* and *Metaxytherium* sp.) differed in having medium and small tusks respectively, whereas the larger bodied form

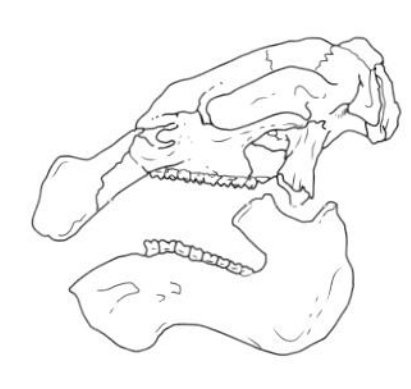

The modern dugong, *Dugong dugon*, is readily distinguished from closely related manatees by a triangular tail used in locomotion and its snout. The dugong's snout is more strongly downturned than the manatees, and reflects the dugong's bottom feeding habit.

(*Dioplotherium manigaulti*) had bigger tusks. The smaller-sized dugongid species were able to feed in shallower waters, and the smaller tusk sizes of these animals suggest that they fed on small plant roots, thereby avoiding food competition with the larger species.

The Indian species all had large tusks but differed in body size and rostral deflection. Two species had similar body sizes (*Domingia sodhae* and *Bharatisiren kachchhensis*) but different rostral deflection. *Domingia* likely uprooted large rhizomes whereas *Bharatisiren* did so to a lesser extent. The smallest of the three species *Kutchisiren cylindrica* showed the greatest rostral deflection, suggesting that it was able to coexist with the others by bottom feeding in very shallow waters.

The Mexican assemblage more closely resembled the Indian species. In the Mexican assemblage two species had large body sizes (*Dioplotherium* sp. and *Corystosiren varguezi*) and another species (*Nanosiren* sp. cf. *N. garciae*) was considerably smaller. Based on tusk size and cranial morphology, there was likely a difference in feeding preference; *Nanosiren* fed on small seagrasses and the two larger bodies species fed on progressively larger seagrasses.

Among the assemblages studied, the Florida species separated by tusk morphology evolved different feeding preferences whereas small body size and strong rostral deflection evolved independently in the Indian and Mexican assemblages. Vélez-Juarbe and colleagues concluded that each multispecies community of sea cows evolved independently from unrelated species of fossil sea cows, yet, each time in different parts of the world, these species separated themselves by utilizing different food resources.

# RICE'S WHALE

## A new species of baleen whale called Rice's whale was unexpectedly discovered living in the Gulf of Mexico.

Despite being some of the largest animals on the planet, it seems somewhat surprising that we are still discovering new species of whales. In 2003, a new species of baleen whale was discovered, called Omura's whale, *Balaenoptera omurai*, inhabiting the Indo-Pacific and the Atlantic Oceans. In 2019, a 38 feet (11.5 m) baleen whale that washed up near the Florida Everglades was thought to belong to yet another new species—Rice's whale (*Balaenoptera ricei*), named for Dale Rice, long-time NOAA marine mammal scientist. In the 1990s, Rice recognized that a small population of whales was living in the northeastern part of the Gulf of Mexico. At the time the assumption was that it was a subpopulation of Bryde's whales (*Balaenoptera edeni*). In 2008, NOAA scientists conducted a genetic analysis of tissue samples from the mysterious Gulf populations. The analysis suggested that the Gulf population was genetically distinct from other Bryde's whale populations.

The "smoking gun" that would add more evidence to the genetic data would be morphological evidence, such as a skull or other diagnostic skeletal elements. Bones of this new whale were not known until 2019 when a carcass was spotted by a fisherman near Sandy Key in the Florida Everglades National Park. Thanks to efforts of the nearby stranding network, scientists were able to observe the skull anatomy in detail and make comparisons with other Bryde's whales. Peculiarities of the arrangement of the skull bones near the blowhole indicated that the whale specimen differed from Bryde's whale. The newly minted Rice's whale is a bit smaller than Bryde's whale, which can be longer than 50 feet (15 m).

Distributional data indicates that Rice's whale primarily occupy the northeastern Gulf of Mexico along the continental shelf break of the De Soto Canyon area at water depths of 490–1,380 feet (150–420 m). Both males and females breed within the Gulf of Mexico. Individuals are mostly seen alone or in pairs but have been observed in larger loose groups believed to be associated with feeding. Their most commonly recorded prey are pelagic schooling fish.

In addition to genetic and morphologic data, Rice's whales have a unique acoustic signature, most commonly a long moan call type that distinguishes them from other species. In a workshop that provided guidelines for cetacean

SIZE

Males & females
Length: up to 41 ft (12.5 m)
Weight: up to 30 tons (27.2 tonnes)

DIET

Pelagic schooling fish

HABITAT

492 ft–1,345 ft (150 m–410 m) deep

LIFE HISTORY

Sexual maturity
Males & females: 9 years
Gestation: 10–12 months
Reproduce: 2–3 years

STATUS

Critically endangered
Threats from pollution, ship collision, and underwater noise

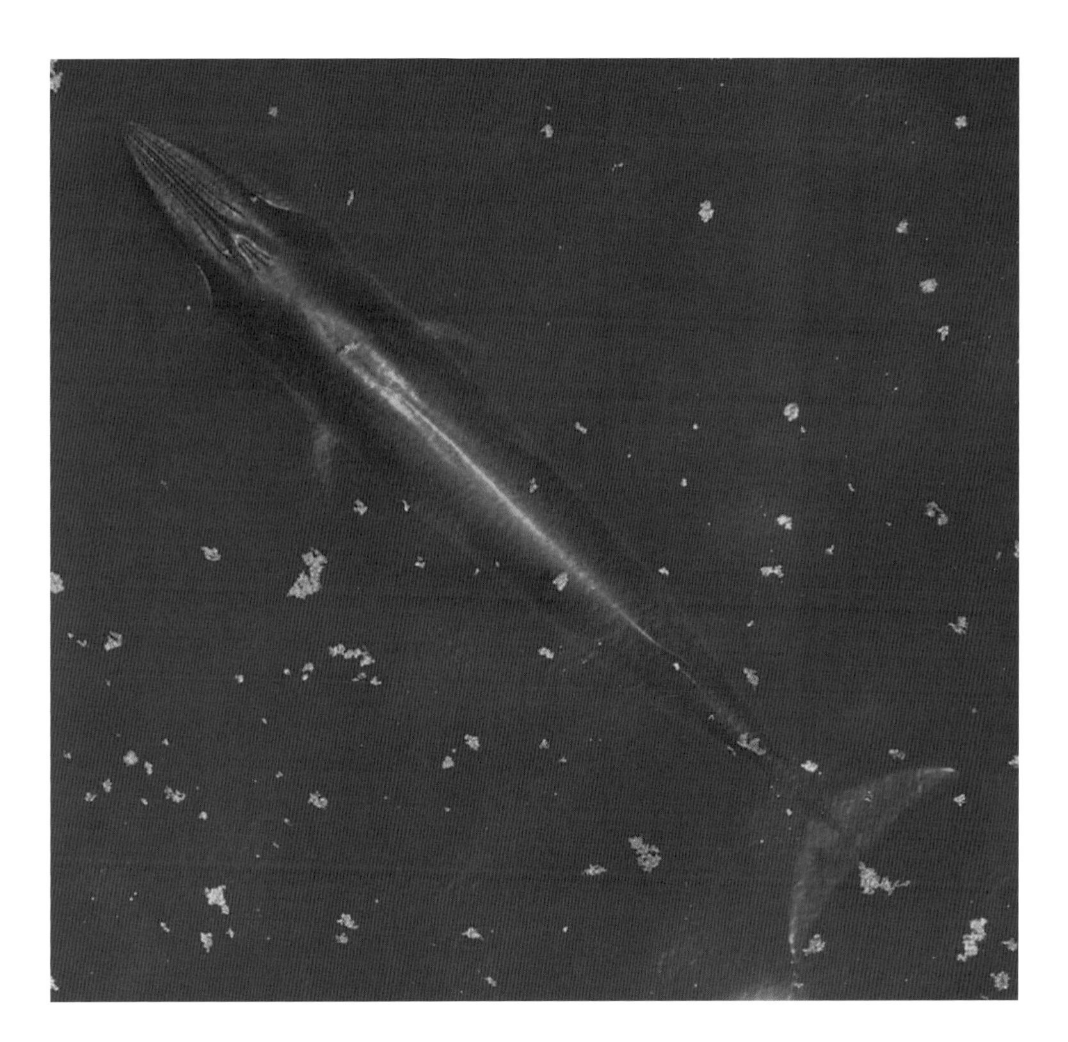

A new baleen whale species, Rice's whale, is located in the Gulf of Mexico and is distinguished from other balaenopterids by its distinctive skull anatomy, and by its call type, which is characterized by a long moan.

taxonomy, it was concluded that two independent lines of evidence were necessary for delimiting species, therefore both morphologic and genetic data support Rice's whale as a new species distinct from other baleen whale species.

The small population size (under 100 individuals) and the restricted distribution of Rice's whale place these whales at high risk of extinction and a cause for grave conservation concern.

Rice's whale is most closely related to Bryde's whale and is a member of the family Balaenopteridae.

# SATO'S BEAKED WHALE

## A newly described species of a rare beaked whale is called Sato's whale. Uniquely, it has a shorter beak and a darker body color than other beaked whales.

Identifying, naming, and describing new species is of critical importance in conservation. Traditionally, there was exclusive reliance on morphological features to distinguish new species, but sometimes morphological differences are very slight or undiagnostic, making the use of another line of evidence such as genetic data important, for example, DNA, RNA sequences.

Beaked whales have the second largest number of species among families of toothed whales, only dolphins (family Delphinidae) are more speciose. They are elusive and seldom seen, preferring deep ocean waters where they feed on bottom-dwelling fish and giant squid. They are thought to have originated and evolved in the Southern Hemisphere. Sightings of the mysterious creature that had a bulbous head and beak-like rostrum had been reported by Japanese fishermen, who call them *karasu* or ravens, for their dark color and small size, but the species was previously unknown to science.

In June 2014 in the Pribilof Islands community of St. George, a biology teacher spotted a 24 feet (7.3 m) long whale carcass on a desolate beach. DNA evidence reported in 2013 supported the possible recognition of a new species of beaked whale although more data was required for verification. The bones of the carcass were DNA analyzed along with those from a whale skeleton hanging from a high school gym in the Aleutian Islands and it was confirmed in 2019 that it was a new species of beaked whale. It was named Sato's beaked whale (*Berardius minimus*) and it lives in the central and western North Pacific (including portions of Japan, Russia, and Alaska) along with the more common and larger Baird's beaked whale. The specific name (*minimus*) reflects the smallest body size of physically mature individuals of this species compared with other *Berardius* species. Other unique characters of Sato's whale are a proportionately shorter beak and a darker body color.

The new species and Baird's beaked whale (*Berardius bairdii*) can be differentiated from Arnoux's beaked whale (*Berardius arnuxii*), which is found in the Southern Ocean. Importantly, study of specimens using multiple lines of evidence—external morphology, osteology, and genetic data—justified naming a new species of beaked whale. Recent observations of groups of Sato's beaked whale in Nemuro Strait between Japan and the Kuril Islands of Russia reveal that this species is shy, lacking visible blows, distinguishing it from *B. bairdii*.

SIZE

Males: length 21 ft–23 ft (6.4 m–7 m); weight unknown
Females: length: 20 ft (6.2 m); weight unknown

DIET

Unknown

HABITAT

Deep ocean

LIFE HISTORY

Unknown

STATUS

Near threatened
Threat from pollution

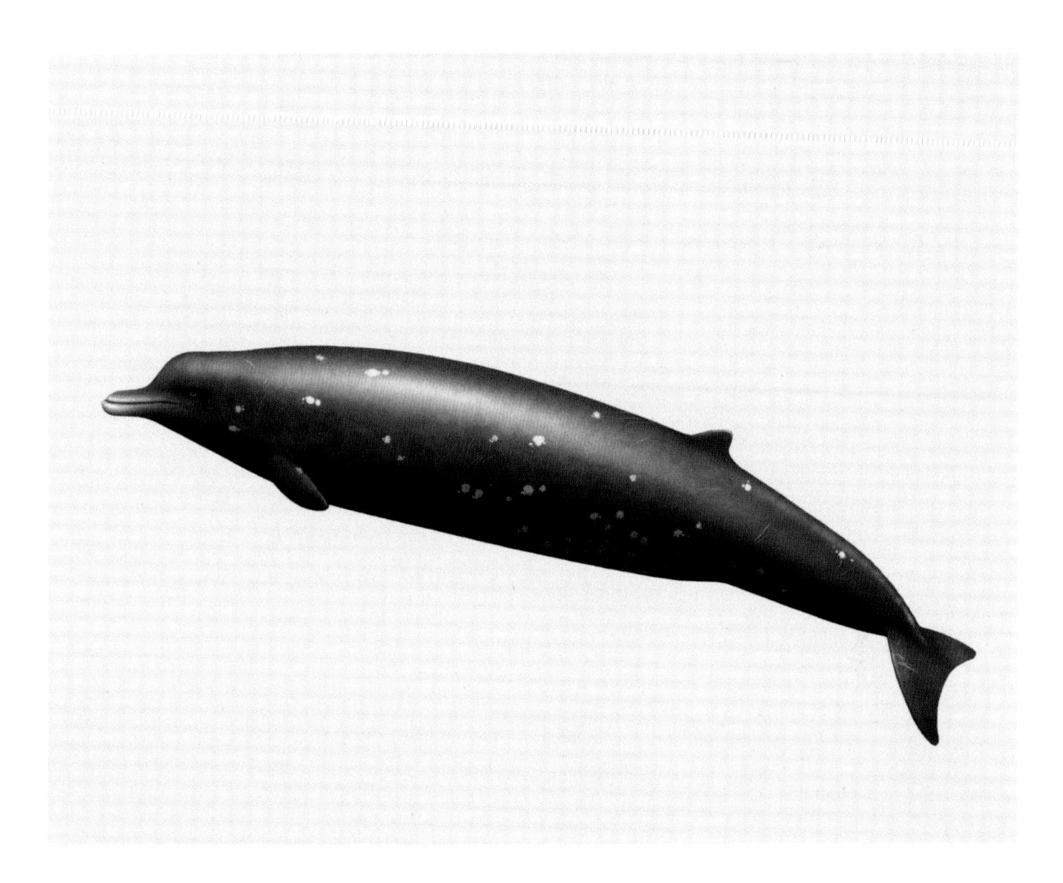

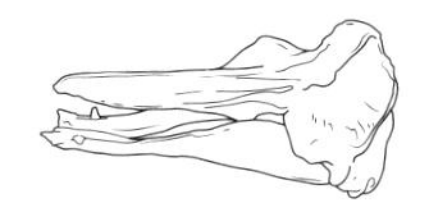

A new beaked whale species, *Berardius minimus*, has a distinctive dark body color. The skull of this species (top) can be differentiated from its close relative Baird's whale (bottom) by its smaller size and shorter beak.

The oldest record of a fossil species of *Berardius*, *B. kobayashii* was described based on a partial skull from Sado Island, Japan, from the middle to late Miocene (12.3–11.5 mya) and provides a minimum age for the origin of the modern genus. Based on the distribution of the fossil and living species of the genus, the western North Pacific including the Sea of Japan may have been one of the areas for the evolution and radiation of this genus at a time before 11 mya. The ancestor of *B. bairdii* and *B. arnuxii* may have evolved somewhere from a *B. kobayashii*-like beaked whale in the Northern Hemisphere or somewhere between the Northern and Southern Hemispheres by the end of the Miocene. This might have been followed by an antitropical dispersal. The two currently extant taxa have overlapping ranges; the smaller *B. minimus* and the larger *B. bairdii* in the North Pacific could correspond to a secondary contact later than the Pliocene. The evidence suggests rapid diversification of *Berardius* species in the initial stage of their evolution during the middle and late Miocene in the North Pacific.

# Fossil Dolphins and Beaked Whales

## Seafloor dredgings have revealed treasure troves of fossil marine mammals, including dolphin and beaked whale species new to science.

Traditional methods of fossil discovery involve excavating bones from a quarry or finding bones poking out of sedimentary rocks. However, fossils sometimes turn up where they were not expected. An alternative method that has proven highly productive involves trawling the deep seafloor in a fishing vessel in search of fossil outcrops. North Sea dredgings of 5–1 mya sediments off the Netherlands coast have yielded an amazing bycatch of fossils. These discoveries contain bones and teeth from various marine mammals including dolphins (delphinids) as well as mammoths, woolly rhinos, hippos, lions, bears, wild horses, bison, elk, reindeer, hyaenas, wolves, sabretooth cats, and our closest extinct relative, a Neanderthal skull. A new species of dolphin dubbed the "balloon-headed dolphin" *Platalearostrum hoekmani* features a short, wide rostrum most similar to short-finned pilot whales, *Globicephala macrorhynchus*. The peculiar morphology of the rostrum and short wide jaws with few teeth may relate to its suction feeding habit.

Compared to other groups, the fossil record of beaked whales is relatively scarce. An unexpectedly large number of well-preserved fossil whale skulls have been recovered from dredgings off the South African seafloors. An exceptionally abundant and diversified association of fossil beaked whales was discovered at depths between 330 and 3,281 feet (100–1,000 m). There are several reasons that help account for this unusual occurrence. First, there has been extensive commercial and scientific deep-sea fishing activities and exploration of the seafloor in this area. Secondly, phosphorites, rocks containing large amounts of phosphate minerals, and which are rich in marine mammal fossils, are abundant in this area. Eight new genera and ten species of fossil beaked whales were recovered. Such a high diversity might be linked to the upwelling system and the resulting high productivity of the Benguela Current, which has been in place and influenced conditions in shallower waters off the coasts of South Africa and Namibia since about 15 mya. A diversity of body sizes is seen among fossil and extant beaked whales. Among living beaked whales this is, in part, due to dietary differences, specifically prey size variability and was likely the case in the past.

Dredgings of the Atlantic seafloor off the coasts of Portugal and Spain have yielded 40 partial skulls of four new species of fossil beaked whales. These new species exhibit extremely bizarre skull morphologies. *Globicetus hiberus* possesses

SIZE

Length: up to 20 ft (6 m)
Weight: Unknown

DIET

Unknown

HABITAT

Unknown

AGE

3–2 mya

The extinct balloon-headed dolphin (*Platalearostrum hoekmani*) had a very short rostrum similar to that seen in pilot whales.

a forehead made of compact bone, somewhat similar to the crests of adult males of the northern bottlenose whale (*Hyperoodon ampulatus*), which are used in head-butting interactions between males. However, in the living bottlenose whale the rostrum is lighter, spongy bone possibly better suited for frontal impacts. The fossil ziphiid's (beaked whale) heavy, dense bone might be related to deep diving, providing ballast to help maintain a vertical position in the water during descent. It is also possible that this unusual facial morphology relates to echolocation. Specimens of *Tusciziphius atlanticus* display conspicuous intraspecific variation especially in the premaxillary bulge on the face, which appears to show sexual dimorphism also seen in several living beaked whales (*Mesoplodon* spp. and *Ziphius cavirostris*). *Imocetus piscatus* bears an oddly elongate facial region and longer sets of maxillary crests. Additionally, significant variation was observed in *I. piscatus* specimens likely influenced by sexual dimorphism. The morphology of the premaxillary and maxillary crests of *T. atlanticus* and *I. piscatus* might protect forehead tissues from lateral impacts or be involved in altering the shape of the high-frequency sound beam produced in the nasal passages. The geologic age of these dredged specimens is uncertain but they are most likely from early to middle Miocene (20–14 mya).

The differences in composition of cold to temperate Northern (North Atlantic) and Southern Hemisphere (South Africa) beaked whale fossil faunas may be explained by a warm water equatorial barrier that kept the two faunas separate.

# Tuskless Walrus

## One of the most completely known fossil walrus skeletons found in southern California was described as a new species of tuskless walrus.

A new archaic walrus *Titanotaria orangensis* was described from the late Miocene (Oso Member of the Capistrano Formation) in Orange County, California. This new walrus is one of the most complete known fossil walrus skeletons in the world, consisting of the skull, jaws, and nearly complete appendicular and axial skeleton. Finding walruses in southern California seems strange but at least 23 known species of walrus assigned to 19 genera, some with two to four tusks and some with none, swam in the waters of Orange County in the past. This fossil walrus was discovered nearly 30 years ago but languished in collections of the John D. Cooper Archaeological and Paleontological Center before it was properly studied by paleontologists James Parham and Isaac Magallanes at the California State University, Fullerton (CSU Fullerton). The walrus was named *Titan*—for the nickname of CSU Fullerton, the Titans, and *Otaria* for the genus *Otaria*, from which other walrus names have been derived. The species name *orangensis* is derived from Orange County, where the type fossil skeleton was found.

Although only a single species of walrus exists today living on shrinking icebergs and habitat in the Arctic, the past diversity of walruses was much greater, occupying seas in the Pacific from Baja to Japan (see page 108). *Titanotaria* is distinguished from the modern walrus and other tusked walruses by its lack of tusks (enlarged canine teeth) and retention of a full complement of upper and lower teeth: three incisors, two canines, four premolars, and two molars. As judged by tooth morphology and wear patterns, the tuskless walrus was a fish eater rather than feeding on benthic mollusks. This species resembles the extinct walrus *Imagotaria* in cranial, dental, and postcranial features, but is larger and more robust. The tuskless walrus is an important addition to our understanding of walrus evolution because it is one of the best-known and latest surviving early walruses.

SIZE

Males & females
Length: over 10 ft (3.3 m)
Weight: 1,200 lb (544 kg)

DIET

Fish

HABITAT

Nearshore

AGE

7–6 mya

The extinct walrus *Titanotaria orangensis* is distinguished by its lack of tusks. Its teeth were well developed and this species was a fish eater rather that a suction feeding mollusk eater like the modern walrus.

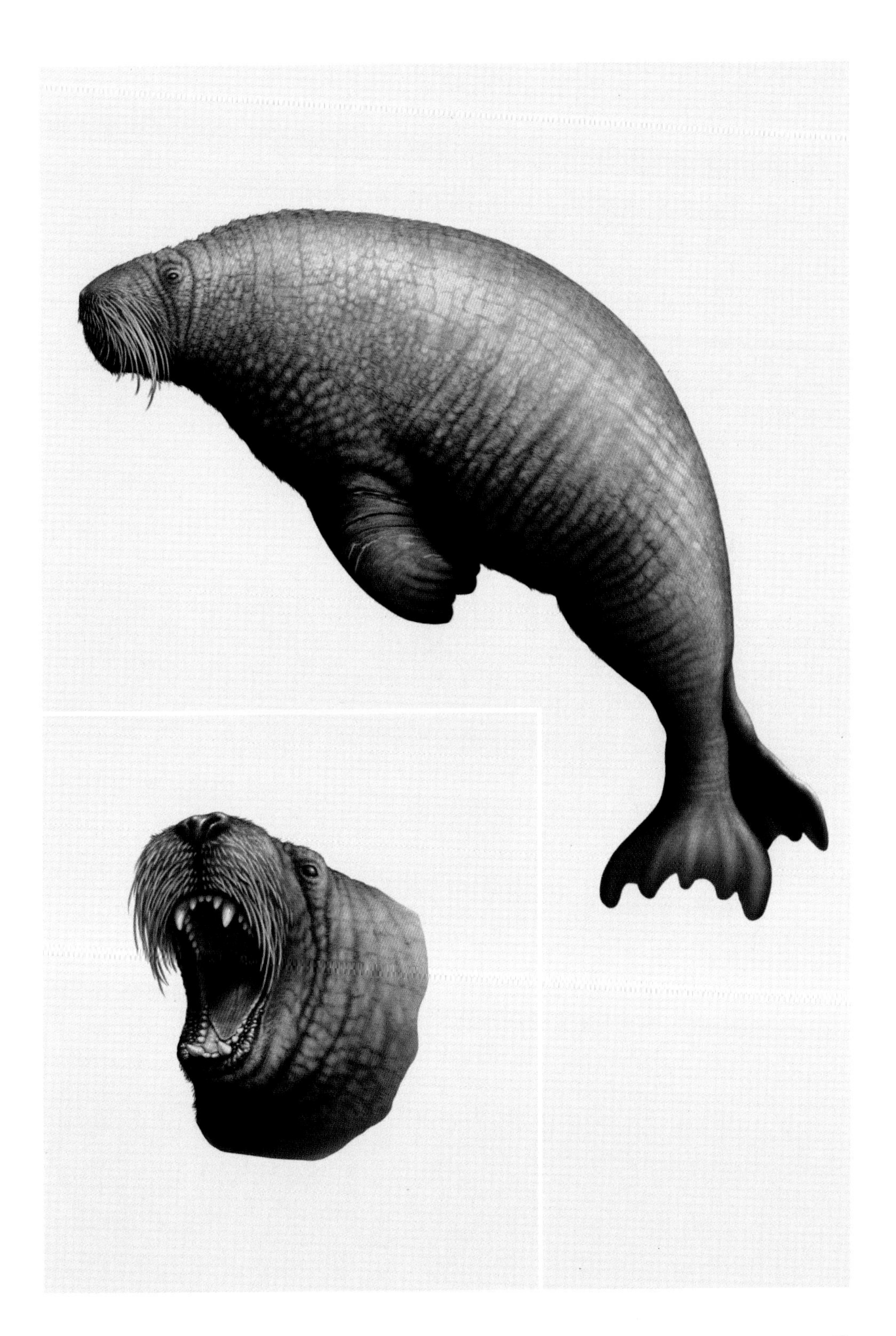

# SEAL RELATIVE

## A partial skeleton of a new fossil seal species described from Washington adds new information about the evolution and extinction of archaic seals.

A new seal *Allodesmus demerei* was described based on a partial articulated skeleton recovered from late Miocene rocks (Montesano Formation) of Washington. This new species has the distinction of being the youngest seal in the family Desmatophocidae coexisting with some other pinnipeds (walruses and otariid fur seals). Several bitemarks and teeth of the dogfish shark were found concentrated around the skull and vertebrae of *A. demerei*, suggesting it was attacked and probably killed by the shark. The large canines and strongly developed nuchal and sagittal crests of the skull indicate that the type and only known specimen of *A. demerei* was an adult male individual.

Although as many as four genera of the subfamily Allodesminae have been recognized only two genera are likely valid—*Allodesmus* and *Atopotarus*. Six species of the genus *Allodesmus* are known from the middle to the late Miocene of the North Pacific based on study of anatomical comparison alongside consideration of age-related changes, sexual dimorphism, and analysis of their evolutionary relationships. All *Allodesmus* species are characterized by having a long narrow skull and very large eye orbits; the latter is thought to be an adaptation for deep diving. *Allodesmus* species display sexual dimorphism, males are larger than females. The single rooted, simple peg-like teeth suggest a fish-eating diet. Thousands of isolated bones of several species of *Allodesmus* have been found at the Miocene Sharktooth Hill deposits in California, suggesting that this animal lived in large social groups. *A. demerei* is most closely related to *A. kernensis* from California. The distribution of *Allodesmus* species suggests either westward dispersal from the eastern to the western North Pacific (Japan) or a diversity of the group that was higher than previously known on both sides of the Pacific. A second subfamily, the Desmatophocinae, consists of a single genus *Desmatophoca* and two species.

Desmatophocidae are an extinct family of seal-like pinnipeds that diversified during the early and middle Miocene. They are the first large-bodied pinnipeds ranging from 4 feet to 11.5 feet (133–350 cm) in length. Desmatophocids were restricted to the Pacific and are known from Baja California, California, Oregon, Washington, and Japan. Their relationship to other pinnipeds is unresolved. They are either most closely related to phocid seals or to otarioids (Otariidae + Odobenidae).

SIZE

Males & females
Length: 8 ft (2.5 m)
Weight: 800 lb (360 kg)

DIET

Fish

HABITAT

Coastal-oceanic

AGE

10.5–9.1 mya

The extinct seal relative *Allodesmus demerei* is characterized by having a long skull and large eyes; the latter is thought to be an adaptation for deep diving.

Desmatophocid diversity coincided with the middle Miocene climatic optimum (17–14.75 mya) and desmatophocids declined as walruses diversified, suggesting competitive replacement by walruses. An alternative explanation proposes that walruses and desmatophocids could have been in competition for haul-out sites for breeding and molting. A third explanation posits an ultimate role for climate change, suggesting that walruses were better adapted in some way to cooler environments. This would have led to decline and extinction of desmatophocids while allowing walruses to increase and diversify. This may relate to a current situation in the North Pacific: "a sequential megafaunal collapse," which has populations of small marine mammals, such as seals, sea lions, and sea otters, decreasing in numbers throughout the twentieth century, victims of changes in the dynamics of ecosystems. Although the causes of these ecosystem changes are not known, two different hypotheses have been proposed. A "top-down effect" argues that a top predator, such as the killer whale today, or fish-eating walruses in the past, were better adapted to environmental conditions and so controlled the ecosystem. The alternative—"bottom-up forcing"—proposes instead that overfishing or decreases in ocean productivity are responsible for the declines in various marine mammal lineages.

# Cerro Ballena

## A fossil whale graveyard was discovered beside an excavation for the Pan-American Highway in Chile and the specimens preserved through 3D imaging.

SIZE

Juveniles and adults

DIET

Krill and fish

HABITAT

Oceanic

AGE

11–6 mya

One of the most astonishing recent fossil discoveries was a death assemblage of whales found in Chile. This graveyard is dominated by nearly 40 complete skeletons of baleen whales, balaenopterids, likely all the same species and encompassed a range of different aged individuals from juveniles to mature adults, all buried belly-up. In addition to baleen whales, at least one species of phocid seal, a species of sperm whale, a walrus-like toothed whale, and an aquatic sloth were found in the Atacama Desert, the world's driest desert. The fossil find, named Cerro Ballena, which means "whale hill" in Spanish, is believed to be the largest fossil mass stranding site. Its discovery occurred when a cutting was made to widen the Pan-American Highway. Scientists at the Smithsonian's National Museum of Natural History were given just two weeks to complete their field work before heavy trucks and equipment were to return to finish construction of the new road. The research team set about recording as much information as possible, using 3D scanning technology to make digital models of the skeletal remains in situ and then removing the bones for further study in the lab. The fossils were found in an area of the roadcut that was just 787 feet (240 m) long and 65 feet and 6 inches (20 m) wide. The local tectonic plates pushed what was once ancient seafloor to more accessible heights, and since the area is so dry, there is little vegetation, and lots of erosion occurring over time revealed the fossils.

Scientists proposed that the marine mammals were poisoned by harmful algal bloom (HAB) toxins, through the ingestion of prey or inhalation, causing relatively rapid death at sea. What makes algal blooms harmful is the presence of algal species that produce toxins like domoic acid, which can cause paralysis and death in mammals and birds. The team noticed that the skeletons were nearly all complete, and that many had come to rest in the same direction, suggesting a catastrophic event(s) and rapid burial. Likely the marine mammals died out at sea and carcasses were thought to have floated toward the coastline in large numbers and hence were rapidly buried without major scavenging or disarticulation, allowing for their exquisite preservation. Some of the whale bones show traces of scavenging by crabs. Likely this occurred initially while the carcasses were floating before burial. The fact that the carcasses occurred in four different fossil levels indicated that it was four separate events that occurred over a relatively short

One of 40 extinct baleen whales likely all the same species found at Cerro Ballena in Chile. This suggests a mass death assemblage perhaps the result of a Harmful Algal Bloom, the likely cause of many repeated whale mass strandings today.

period of 10,000 to 16,000 years, a snapshot of time, in the late Miocene epoch (11–6 mya). Although the road expansion is now complete and the place where many of the fossils is paved over, fossils found between 2010 and 2013 have been moved to museums in Chile and await preparation. Cerro Ballena potentially contains many more fossils, perhaps several hundred fossil marine mammal skeletons, as yet undiscovered.

The conditions that lead to an HAB-mediated mass stranding are tied to upwelling systems along continental margins, consistent with the hypothesized occurrence along the South American coastline. Such blooms are one of the prevalent causes for repeated mass strandings seen today. Although mass mortality events (MMEs) of marine mammals typically involve social species such as dolphins, pilot whales, or sea lions, they do occur in baleen whales that exhibit less gregarious behavior. In 1987, for example, 14 humpback whales became stranded along the coastline in Cape Cod over a span of five weeks. Like the Cerro Ballena fossils, the stranded whales included males, females, and calves. When scientists cut them open they found Atlantic mackerel in their bellies and the mackerel were loaded with algal poisons. All the signs at Cerro Ballena suggest that these animals died in a similar way.

## Causes of Mass Marine Death

Other fossil marine mammal death assemblages have different explanations. For example, the Miocene Pisco Formation of Peru, the site of numerous well-preserved fossil sharks and rays, bony fish, seabirds, turtles, crocodylians, aquatic sloth, seals, and whales (see page 24) did not have a single cause. It is likely there were a combination of causes, ranging from original biological abundance to low oxygen levels at the seafloor and rapid burial, which together created conditions favorable to fossilization.

A different cause for another marine mammal death assemblage is proposed. The middle Miocene Sharktooth death assemblage in the Round Mountain Silt exposed near Bakersfield, California, is one of the most significant, diverse accumulations of sharks, whales, seals, and a seal relative. It has been hypothesized as the result of a sudden mass stranding or red tide poisoning, or the remains of a calving ground for marine mammals. Instead, it is now believed to be the result of slow, steady bone accumulation during a long period of time when there was little sand, silt, and mud sedimentation. Currents swept away the bone beds over several hundreds of thousands of years during which time the bones gathered in a big and shifting pile.

Another mass marine death assemblage is at Wadi Al Hitan "Valley of the Whales," a UNESCO Natural Heritage site in the Western Desert of Egypt, where more than 400 whale and sea cow skeletons (see page 14) as well as sharks, body fish, crocodylians, and sea turtles have been found. The site spans more than 6 miles (10 km), deposited during the middle Eocene about 40 mya. The deposits contain well-preserved, virtually complete articulated skeletons, some with stomach contents preserved. The presence of juvenile skeletons suggests that the area was a shallow bay, perhaps a calving ground for various marine species.

OPPOSITE TOP
3D model of a baleen whale skeleton that was found at Cerro Ballena.

OPPOSITE BOTTOM
The excavation of three adult and juvenile (center right) baleen whale skeletons located at Cerro Ballena.

# 3
# BIOLOGY

Stories of the biology of ten living sea mammal species are told. These species were chosen for their adaptations to living in water, including unique aspects of anatomy, habitat, communication, and longevity.

Baleen whales filter plankton through plates of whalebone (baleen) in their mouths. Two baleen whales have made an amazing comeback after being hunted to near-extinction during commercial whaling. Blue whales are awe-inspiring marine giants with impressively big mouths enabling them to efficiently feed on the oceans' smallest organisms. Humpback whales use their long, graceful flippers to expertly maneuver in shallow water, as well as serving to advertise their presence when forcefully slapped against the water. Another baleen whale, the rotund bowhead, is distinguished by being the longest-lived mammal. Its higher resistance to aging and some diseases make it a valuable study animal with implications for human health. These, as yet, unstudied preventative mechanisms and genes in bowhead whales may confer resistance to cancer and delay the aging of cells in humans.

Toothed whales include the killer whale—the largest dolphin—an alpha predator at the top of the food chain, and at least six different killer whale types varying in size, appearance,

PAGE 84
The Amazon river dolphin is also known as the pink river dolphin. It is the largest of the river dolphins.

prey preferences, foraging techniques, dialects, behaviors, and social groups. These ecotypes are probably distinct subspecies, or perhaps even species in the making. The Amazon river dolphin, the best known of the river dolphins, is the only species with a pink coloration due to the temperature and mineral context of the freshwater lakes and rivers it occupies in South America.

Pinniped species profiled include the hooded seal named for the stretchy cavity or hood in the nose of male seals, which when inflated resembles a red balloon that attracts females during the mating season, and shows aggression to other males. The walrus is distinguished by its elongate tusks, which are in fact teeth. The tusks are displayed to signify maleness rather than employed as a foraging tool. The ringed seal has a coat conspicuously marked by white rings that encircle dark spots, the coat functions for species recognition in the stark, white Arctic background. Their long, sensitive whiskers can sense sound waves underwater, enabling them to navigate and to discriminate prey size and shape.

# BLUE WHALE

## The blue whale has the distinction of being the largest animal alive today, as well as being the biggest creature to have ever lived, rivaled only by a few dinosaurs.

Blue whales (*Balaenoptera musculus*) get their name from the bluish gray coloration on their back that is lighter on their bellies. They are slender bodied and streamlined with females larger than males. Blue whales have a cosmopolitan distribution, occupying all the world's oceans except the Arctic. Four subspecies are recognized, based on geographic differences.

Blue whales are usually seen alone or in pairs. Larger aggregations are sometimes seen on feeding grounds. Some are resident but most undergo long distance migrations, moving poleward to feed in the summer and toward the tropics to breed in the winter.

The large size of blue whales is possible, in part, because they live in the ocean which enables them to use buoyancy to stay afloat in water, thereby countering the effects of gravity. But there is more to the story.

Blue whales and their kin are also large. Gigantism in baleen whales is tied to prey choice and the coincidence of their evolution with a global increase in the upwelling of nutrient-rich ocean water. Lunge feeding evolved in baleen whales about 10–7 mya, which enabled them to swim through the ocean and engulf huge amounts of water containing large concentrations of plankton. A unique sensory organ has developed among blue whales and their kin, embedded within the connection between the lower jaws, which consists of mechanoreceptors that help the brain to coordinate lunge-feeding events.

Although the first baleen whales (35 mya) used their teeth to feed on fish, they later lost teeth and evolved baleen, keratinized plates that hang in racks from the upper jaw that allows them to filter feed on large batches of prey (see page 26). The mouth of blue whales contains 300–400 baleen plates that are moderate in length, less than 3 feet (1 m) long with coarse bristles.

Scientists recently reported the fossil remains of a blue whale found in Italy approximately 85 feet (26 m) in length from sediments 1.5 mya. This information together with size data from other baleen whales suggests that the greater size of baleen whales occurred 3.6 mya, or perhaps even earlier. Large-bodied baleen whales, however, were not always dominant. Relatively small species were the norm throughout most of baleen whale history.

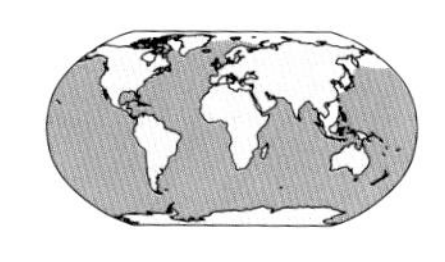

SIZE

Males & females
Length: 75 ft–95 ft (23 m–29 m)
Weight: 79 tons–149 tons (72 tonnes–135 tonnes)
Females larger than males at birth: 23 ft–26 ft (7 m–8 m)

DIET

Various species of krill

HABITAT

Offshore, tropical to polar waters, migratory

LIFE HISTORY

Sexual maturity
Males & females: 7–12 years
Gestation: 1 year
Reproduce: 2–3 years

STATUS

Endangered
Threat from ship collision; in the past commercially hunted

A blue whale dives below the surface. Blue whales can dive to depths of more than 300 feet (100 meters).

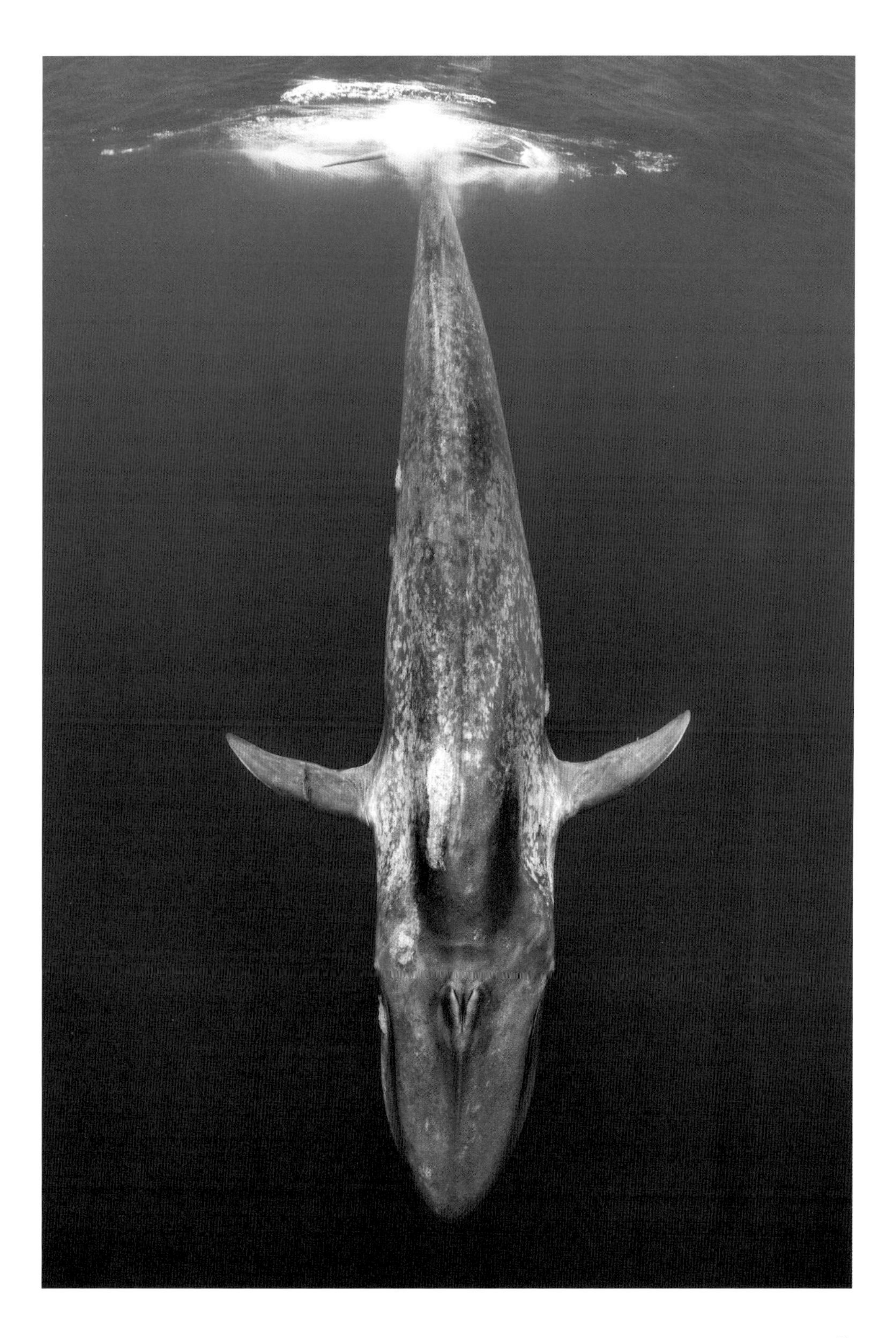

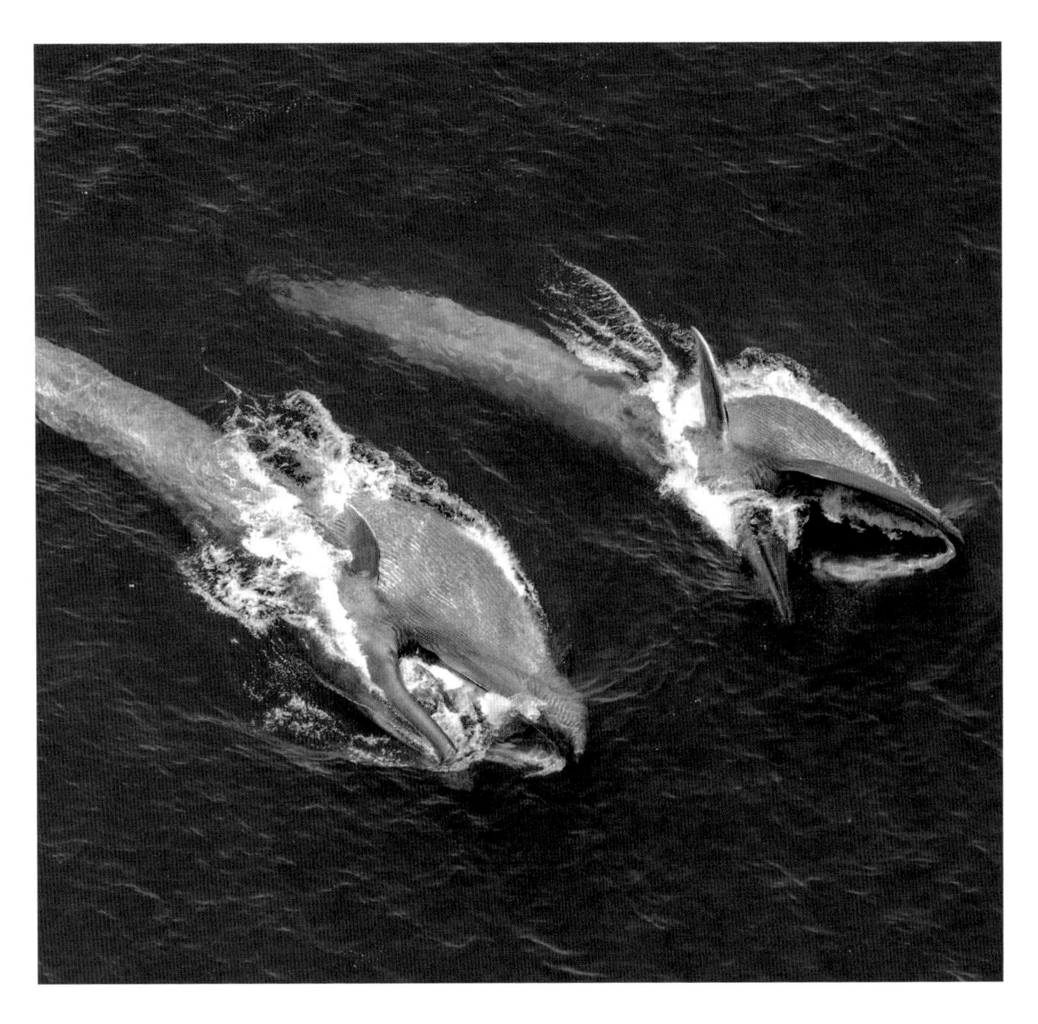

When blue whales lunge feed their throats inflate.

## WHY ARE BLUE WHALES LARGER THAN THE OTHER BALEEN WHALES?

The key to their size and spectacular success is their specialization to feed on krill—tiny shrimplike crustaceans. Krill are abundant in certain areas of the world's oceans, such as upwelling zones, along continental margins, and in polar oceans. Upwelling is driven by wind that pushes surface water away from the shore, which causes deeper water laden with prey to rise to the surface. To survive on massive amounts of krill, blue whales need to have enormous mouths to go along with their huge bodies. A blue whale's mouth is so large that it can engulf a mouthful of water equivalent to its body mass. Pleats of skin below its mouth, known as throat grooves, extend to its belly forming a vast, cavernous extension of its mouth.

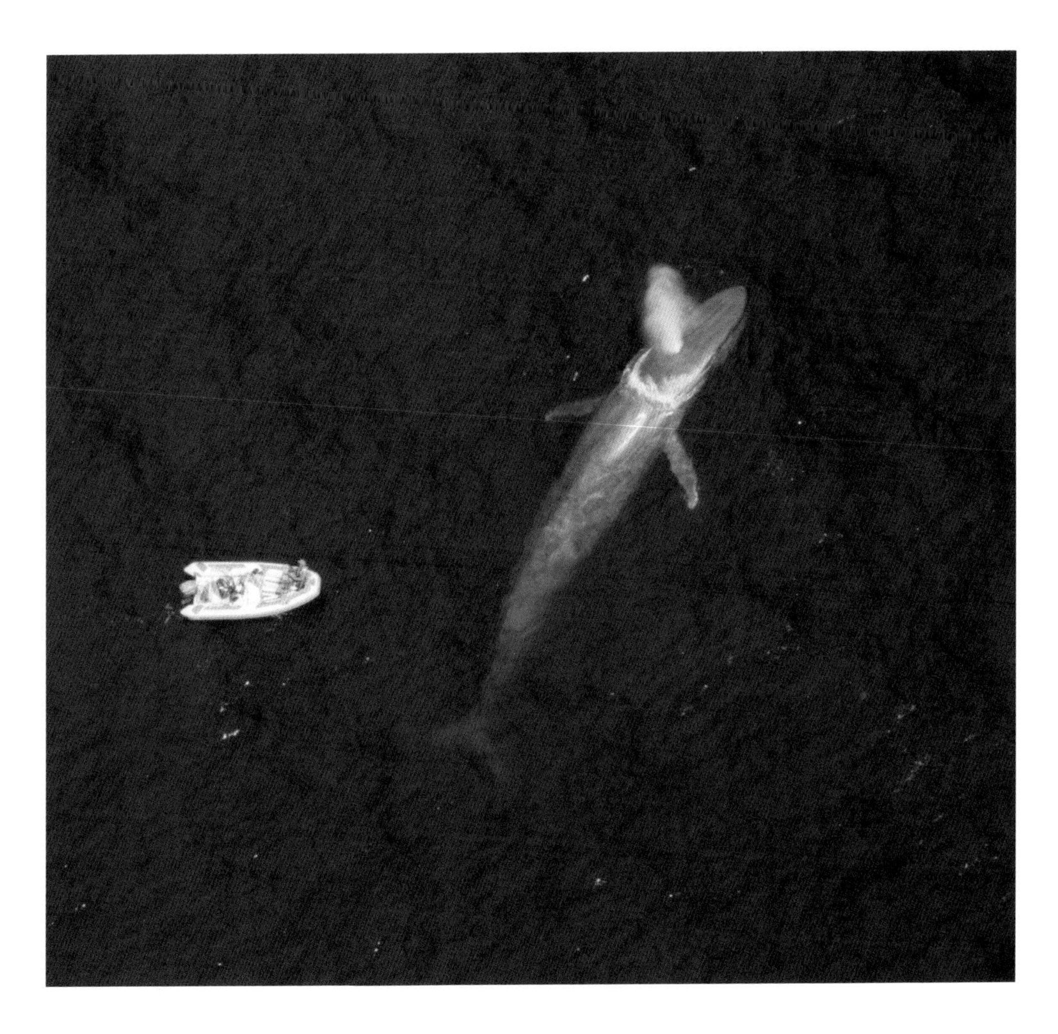

The huge size of the blue whale is illustrated here, as a researcher in a small boat attempts to attach a satellite tag.

Despite surviving commercial whaling and being hunted to near extinction (only 0.1 percent were left in the Southern Ocean), the food supply of blue whales is at risk today, due in part to climate change and overfishing. Moreover, they are also subject to other threats, such as shipping noise, collisions, and pollution. Whereas some baleen whales, such as humpbacks and minke, seem to adapt to changing conditions and can switch prey in order to feed on different fish, blue whale are krill specialists whose only adaptive response has been to get larger. Unless blue whales can become generalists and adapt to declining krill productivity they have "nowhere to go and no way to grow." Climate change could very well determine if blue whales stay at their massive, awe-inspiring size, shrink down, or die out.

Blue whales are one of nine species of baleen whale in the Balaenopteridae family. They are most closely related to the sei whale and Bryde's whale, although their exact placement is still debated.

# Killer Whale

## Killer whales or orcas are the largest members of the dolphin family and among the most widely distributed mammals, occupying every ocean basin from the polar regions to the Equator.

Orcas are easily recognized by their black and white coloring and their tall dorsal fin. Although only a single species of killer whale is recognized at present, *Orcinus orca*, as many as 11 different geographic forms of killer whales have been identified in the field, based mainly on their striking coloration, especially the size, shape, and orientation of the eye patch, and presence or absence of a dorsal cape. Apart from physical differences, each of these different types of killer whales have specific language, behaviors, habitat, and diet. The taxonomic status of these forms is unresolved, but evidence suggests that more than several species and subspecies are present.

Killer whales are incredibly social, intelligent, and powerful predators. Their diet ranges from squid to great white sharks to other marine mammals, including sea otters, seals, sea lions, and whales. Orcas employ a diversity of feeding strategies. They work together to cooperatively herd schools of fish. They also beach themselves to take seals and sea lions on land, or create waves to dislodge seals and penguins from ice floes. Studies in the eastern North Pacific from Washington to Alaska distinguish three ecotypes of killer whales that show prey preferences, such as the primarily fish-eating resident killer whales in the Pacific Northwest (so-called because they remain in inland or nearby coastal waters); or Bigg's (transient) killer whale (referring to a migratory population occurring in nearshore and pelagic waters) that hunt large mammal prey and offshores (with extensive ranges) that appear to be shark specialists.

Feeding on marine mammals by killer whales and false killer whales likely evolved independently less than 1.3 mya in two different dolphin lineages. Multiple lines of evidence including prey remains, skull morphology, tooth wear, and body size show that the ancestors of both false killer whales and killer whales had a fish-based diet. Modern killer whales appear to have expanded their initial fish diet to include new warm-blooded prey. Taking advantage of ecological opportunity allowed both killer whales and false killer whales to feed on multiple trophic levels and adapt to changing environmental conditions.

Orcas are among the best studied cetacean species with several long-term studies spanning more than three decades, such as British Columbia and Washington populations. Orcas live in family groups called pods, which have

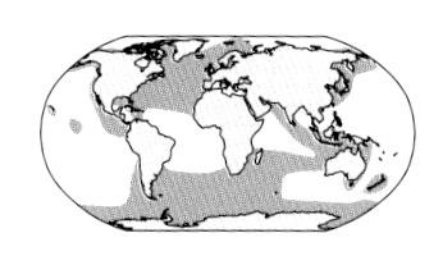

SIZE

Males: length up to 32 ft (9.8 m); weight 11 tons (10 tonnes)
Females: length up to 27 ft (8.5 m); weight 8 tons (7.5 tonnes)
At birth: length 7 ft–9 ft (2.1m–2.8 m); weight 353 lb–397 lb (160 kg–180 kg)

DIET

Fish (sharks, rays, bony fish), other mammals, seabirds

HABITAT

Oceanic and coastal

LIFE HISTORY

Sexual maturity
Males: 15 years
Females: 10–15 years
Gestation: 15–18 months
Reproduce: 5 years

STATUS

Data deficient
Threats include disturbance from ship traffic, lack of prey

The bold black and white coloration of the orca, as well as the tall dorsal fins, make it easily recognizable.

up to 50 members. The pods are matrilineal and made up of related mothers and their offspring and the offspring of their daughters. The primary social unit, the matriline may include as many as five generations of matrilineally related individuals. A male orca stays with its mother for life, while daughters may move to other pods. Differences among overlapping populations, such as in vocal dialect, help prevent inbreeding.

Despite their name, there is no record of a wild orca killing a human. Orcas in captivity, however, have attacked and killed people. While not intentionally harming people, orcas have attacked boats, doing so with more regularity in recent years. These attacks are mostly by juvenile males. The cause is unknown but explanations range from killer whales lacking food to defensive measures the orcas adopted to protect themselves against boat injuries.

RIGHT
An orca hunts South American sea lions on a beach at Punta Norte, Valdes Peninsula, Argentina. This hunting strategy involves orcas chasing seals up onto beaches.

OPPOSITE LEFT
A pod of orcas swim in water located off Tromsø, Norway.

## FISHING UP THE FOOD CHAIN

As the top predators in the oceans, orcas are important members of ecosystems. Killer whale predation is one of the major factors driving the decline of numerous populations of marine mammal species in the North Pacific. In a further complication, killer whales are declining in numbers where northern sea lions (western populations) and sea otters (especially Alaskan populations) are in crisis, and abundant where northern sea lions are showing signs of recovery. An extension of this is that after peak commercial whale hunting, killer whales no longer had the large whales (baleen and sperm whales) as a food source, so they prey shifted to feed on northern sea lions. This has been viewed as a "top-down effect," which refers to a top predator, in this case the killer whale, controlling the ecosystem. Following the decline of the western population of northern sea lions, killer whales continued to prey switch, more recently hunting harbor seals, northern fur seals, and sea otters.

A study of migration patterns of whales in the Arctic revealed another shift in predator-prey relationships, this one more directly the result of climate change. Killer whales were found spending longer in the Arctic Ocean in recent years despite the risk of ice entrapment and they seemed to follow the decrease of sea ice in areas. This suggests that reduction in sea ice may be opening up new hunting opportunities for killer whales, if certain species of prey can no longer use the ice to avoid this predator. Another study provided evidence that this was the case, documenting that bowhead whales, which typically use the ice to hide under, were attacked by killer whales. Killer whales off the southwestern coast of Australia have been observed successfully hunting even larger prey, blue whales, a case of the biggest predator taking down the biggest prey.

# BOWHEAD WHALE

## A unique feature of bowhead whales is that they get astonishingly ancient—up to 200 years old—and are among the longest-living animals, based on the recovery of stone harpoon tips in harvested animals.

Bowheads (*Balaena mysticetus*) are named after their unique, steeply arched upper jaw and mouth that are shaped like an archer's bow. They have a dark, rotund body and unlike most cetaceans lack a dorsal fin. Their large, thick skulls are used to break through ice more than 8 inches (20 cm) thick. Their immense head can be up to two-fifths of their body length. They are skim feeders, swimming slowly through the water with open mouths. The baleen of bowheads is longer than that of other baleen whales, up to 13 feet (3.9 m) long, and they are specialized for feeding on copepods and krill.

Bowheads show few signs of the age-related diseases that plague other animals, including humans, which means their genes are of particular interest. The bowhead genome has been sequenced and compared with the genomes of other mammals. Despite having over 1,000 times more cells and thus putting them at greater risk of developing cancer, results indicate that they don't develop cancer. Mutations in two genes that are thought to confer resistance to cancer are also linked to aging and DNA repair. These longevity genes could lead to drug therapies that fight aging and ultimately help humans live longer.

Bowheads are very vocal and have a large variety of calls. Like other baleen whales they are specialized to produce and hear low-frequency sounds, which are capable of traveling great distances in the ocean and allow long-distance communication.

Bowheads spend their entire lives in Arctic waters in the Northern Hemisphere, from Alaska to Greenland, and they have evolved blubber that is up to 1.6 feet (49 cm) thick to keep warm. They are closely associated with ice for most of the year, and are capable of breaking ice that is 8–24 inches (20–60 cm) thick. They are slow swimmers and live in mostly shallow, coastal water, migrating further north during the summer as the ice melts. Bowheads are usually seen in small groups of three or fewer, but larger aggregations are seen during migrations and on the feeding grounds. Similar to right whales, mating groups usually consist of the female and multiple males courting them.

Bowheads are one of four species of whales in the family Balaenidae. Their closest relatives are the right whales.

SIZE
Males: length up to 59 ft (18 m); weight: 60 tons (54 tonnes)
Females: length up to 66 ft (20 m); weight heavier than males

DIET
Mostly copepods

HABITAT
Oceanic and migratory

LIFE HISTORY
Sexual maturity
Males: 20 years; earlier for females
Gestation: 12–16 months
Reproduce: 3–4 years

STATUS
Least concern
Threat from climate change; in the past commercially hunted

RIGHT
The head of a bowhead whale reveals the steeply arched upper jaw and mouth.

OVERLEAF
Group of bowhead whales socializing in shallow water in the Sea of Okhotsk, Russia.

# Humpback Whale

## Humpback whales with their slender bodies, elongate flippers, and large flukes are one of the most iconic and best-known baleen whales.

The name "humpback" comes from the distinctive hump on the whale's back and its extremely long flippers up to 16 feet (5 m) in length, which inspired its scientific name *Megaptera*, meaning "big-winged." The ventral side of the flukes varies from all black to all white, with individually distinctive pattens of white and black. This fluke patterning is used to identify individual humpback whales (*Megaptera novaeangliae*), known as photo identification, and catalog occurrences of the photo ids allow individual whales to be tracked over time. As is true for most baleen whales, females are larger than males.

Humpbacks live in all the world's oceans and can travel great distances during their seasonal migrations, alternating between cold, nutrient-rich high latitude waters during summer months and warm, shallow calving grounds during the winter. Humpbacks make some of the longest migrations, up to nearly 5,000 miles (8,000 km) one way. Four populations of humpback whales occur in the North Pacific, two populations in the North Atlantic, with seven populations found in the Southern Hemisphere. All populations feed in polar or subpolar regions of Antarctica. Humpbacks generally occur singly or in small groups, although larger groups are present on feeding and breeding grounds. The only long-term bonds are those of mothers and calves. On the breeding grounds, males compete for access to females, using songs as well as physical confrontations that are sometimes violent.

Humpbacks are the most acrobatic of the great whales and they are a favorite of whale watchers. They exhibit an array of behaviors, including jumping out of the water or breaching, as well as slapping the surface of the water with their flippers. These displays occur at all times of year and in a variety of contexts and for a range of functions including play, communication, and parasite removal. The leading edge of the flippers uniquely possess a series of bumps or tubercles, which give the flipper a scalloped appearance. These tubercles reduce drag, increase lift, and allow maneuverability.

Humpback whales have been extensively studied. They are lunge feeders swimming rapidly through water and engulfing enormous mouthfuls of water containing small crustaceans (mostly krill) and fish that are strained through baleen plates. Although they often feed singly or in small groups, humpbacks sometimes gather into coordinated groups of up to 20 or more whales and work

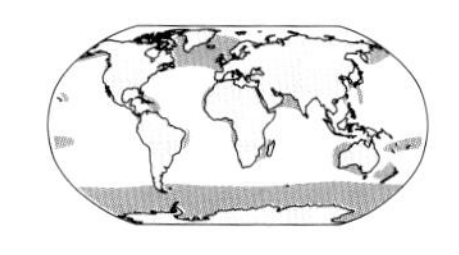

SIZE

Males: length 36 ft–56 ft (11 m–17 m); weight slightly less than females
Females: length 39 ft–61 ft (12 m–18.5 m); weight 35 tons (32 tonnes)

DIET

Mostly krill and small schooling fish (herring, sand lance, sardines)

HABITAT

Coastal, migratory

LIFE HISTORY

Sexual maturity
Males & females: 8 years
Gestation: 11.5–12 months
Reproduce: 2 years

STATUS

Least concern
Threats from fisheries' gear entanglement, ship collision; in the past commercially hunted

A young humpback whale breaching in Frederik Sound, Alaska. Breaching is thought to help humpbacks communicate and to remove parasites from their bodies.

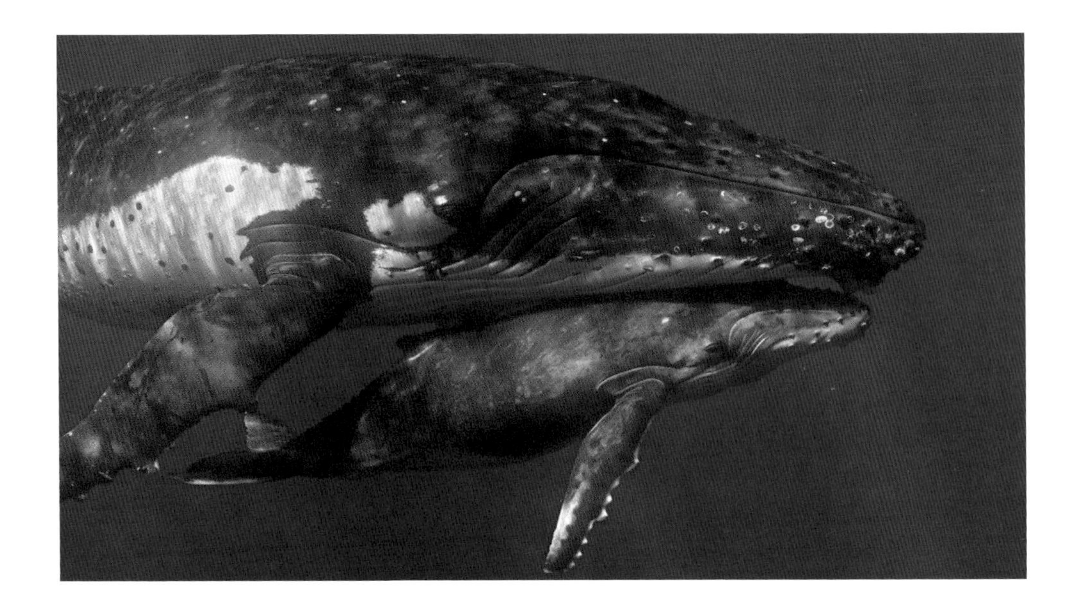

A humpback mother protects her baby beneath her body. Notice the extremely long flippers characteristic of humpbacks with a series of tubercles on the leading edge that help reduce drag.

together to capture prey, often deploying bubbles to herd, corral, and disorient schooling fish, known as "bubble net" feeding. A recent study that used high-tech tagging devices to record movements of whales found that the krill-feeding humpbacks and blue whales in the Southern Ocean eat two or three times as much as previously thought. The reason whales need to eat more krill is to get the same energy as small fish.

However, how do humpbacks (and other baleen whales) find food? Although the cues they use to find food are largely unknown, there is some behavioral evidence that suggests that humpbacks and probably other baleen whales perceive chemical cues, known as chemoreception, especially odors from prey. Some planktonic prey, such as krill, give off specific odors, such as dimethyl sulfide (DMS), released in areas of high marine productivity. Field experiments showed that humpbacks were attracted to chemical sources including DMS. Other marine predators such as sharks, bony fish, sea turtles, seals, and oceanic seabirds have been previously shown to partly rely on chemoreception to locate food.

Humpback whales were heavily exploited by the whaling industry, the result of their coastal distribution. But most populations are making a strong recovery and increasing in numbers today.

Humpback whales are one of seven species of the family Balaenopteridae, the largest group of baleen whales, and they are most closely related to gray whales (family Eschrictiidae).

## SINGING WHALES

Humpbacks are among the most sonorous of the baleen whales, possessing an astonishing vocal repertoire including songs, feeding calls, and social sounds. The males broadcast beautiful, acoustically complex songs that are shared by all singing whales occupying the same breeding ground. Observations suggest that humpback songs play a role similar to that of bird songs—to communicate with others and to attract mates. In part, because of the prominent male display to attract females, the mating system of humpbacks has been compared to a lek, also seen in some dugongs (see page 163).

New research shows that local geography determines the pattern of song transmission that develops. While humpbacks in the Northern Hemisphere have long been known to sing songs that slowly evolve over decades, previous studies of whales in the Southern Pacific revealed songs undergoing dramatic cultural changes, with males rapidly replacing the song of the previous year with an entirely new song adopted from a neighboring population. Southern Hemisphere humpbacks aggregate around feeding territories in Antarctica. This means that interactions between populations are rare but, in these rare encounters, new songs heard from a neighboring group can be learned, then rapidly spread within that population. In contrast, both oceans in the Northern Hemisphere are constrained by continents on both sides, funneling whales of several populations into comparatively small areas. This means that the more frequent interactions in the north create slower changes to a communal song in that ocean basin. With these high rates of interactions, new variants and modifications of songs in the Northern Pacific—as opposed to wholly new songs in the south—are rapidly shared between all populations, leading to one slowly evolving song type dominating the entire ocean basin.

# HOODED SEAL

## Hooded seals are large, deep-diving northern seals famous for their balloon-like "hood" that live associated with pack ice in the cold waters of the Arctic and North Atlantic Oceans.

Hooded seals (*Cystophora cristata*) are sexually dimorphic with the males larger. Adults are white with numerous small to large, irregularly shaped dark blotches, which make them easily recognizable on the ice to other hooded seals. Pups called "blue backs" are born with a coat of dark blue-gray above and creamy white below known as countershading. This helps to camouflage them when in the water since the lighter color on the belly blends in with light from the ocean's surface when viewed from below. This coloration is retained for up to two years after which the spotted adult pelage begins to appear. It is this luxuriant pelt that was once the most valuable to the sealing industry during the 1800s. In addition to the pelt, hooded seals have been harvested for meat and blubber.

Hooded seals have a polygynous mating system in which a few males monopolize mating. Males generally do not breed until they are larger and able to compete with other males. Males are territorial and patrol ice margins, often hauling out near females and forming trios.

Females are usually spaced widely apart and aggressively defend their pups. Hooded seals breed and give birth on pack ice and have a short breeding season, generally only a few weeks in length. They have the shortest nursing period of any mammal, from 5 to 12 days. During this period the pups can double in weight. The demands of nursing, fasting (due to lack of food), the harsh polar climate, and the instability of pack ice all favor a short nursing period. Immediately after the pup is weaned mating occurs. Apart from breeding and molting, hooded seals form loose aggregations. Despite their migratory habits all hooded seal populations share the same genetic diversity that suggests that populations intermingle and mate.

Hooded seals typically fast during breeding and molting but the rest of the year they feed on a variety of marine prey, especially fish. They also feed on octopus and shrimp. In the fall and winter hooded seals feed more on squid and switch to primarily fish in the summer.

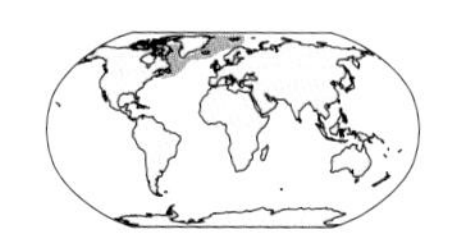

SIZE

Males: length up to 8 ft 6 in (2.6 m); weight 423 lb–798 lb (192 kg–352 kg)
Females: length 6 ft 6 in (2 m); weight 320 lb–661 lb (145 kg–300 kg)

DIET

Squid and fish (cod, halibut, herring, capelin)

HABITAT

Coastal—deep water, pack ice

LIFE HISTORY

Sexual maturity
Males: 4–6 years
Females: 3–5 years
Gestation: 12 months
Reproduce: yearly

STATUS

Vulnerable
Threats from fisheries bycatch, subsistence hunting; in the past commercially hunted

ABOVE
A male and a female hooded seal along with a four-day old pup form a triad on pack ice in the Magdalen Islands, Canada.

RIGHT
Female hooded seal searching for prey.

Hooded seals are expert divers, second only to elephant seals in their diving ability. They can dive for up to one hour and reach depths exceeding 5,249 feet (ca. 1,600 m) although they more regularly dive for 15 to 20 minutes to depths of not more than 1,640 ft (500 m). Hooded seals dive mainly to feed although like other deep-diving seals they carry out "drift dives," during which they descend rapidly with the rate of descent slowing abruptly for 10 to 15 minutes when the seal remains motionless, and appears to be resting, thus reducing energy expenditure. The seal then swims rapidly to the surface. It is thought that seals use the drift dive to sleep and/or digest food.

OPPOSITE
Adult male hooded seals engaged in an aggressive encounter in the Magdalen Islands, Canada.

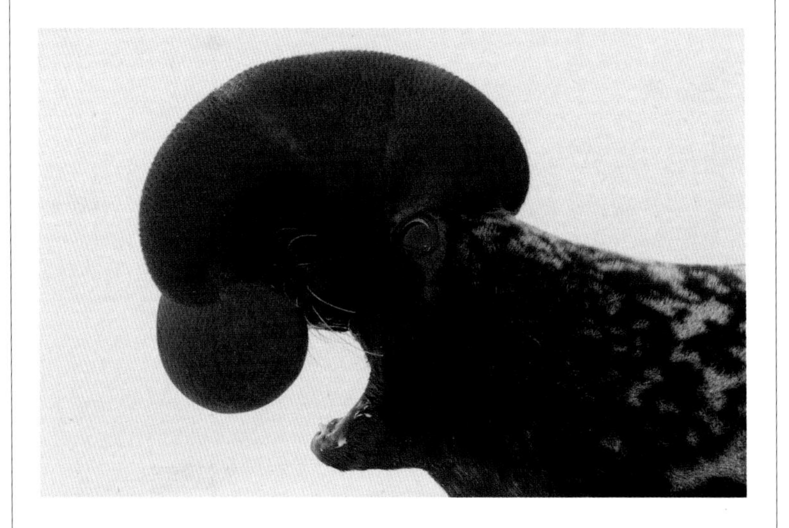

## ORIGIN OF THE "HOODIE"

Male hooded seal heads are inflatable. While that's not quite true, a sac above their nose can be blown up, forming a red balloon covering their nose like a hood. This distinctive physical feature develops at four years of age. It's an extravagant sight that males use to attract females and possibly threaten other males. Hooded seals are very aggressive compared to other seals. When the seal is relaxed, the hood hangs in a loose, wrinkled sac over the front of the nose. When the balloon is inflated, the male shakes it violently and also vocalizes at the same time, making a loud "pinging" sound. If acoustic and visual displays do not discourage other males, a male hooded seal will resort to open combat—biting, pushing, and clawing each other. These encounters usually result in a male victor on the ice near the female and her pup. These temporary social units, termed "triads," are widely separated from each other on the ice.

# WALRUS

## The presence of tusks in the walrus are the most distinguishing features of this iconic Arctic pinniped and are mainly used in male dominance battles.

The modern walrus (*Odobenus rosmarus*), the largest Arctic carnivorous aquatic mammal, is known by a single species placed in its own family of pinnipeds, Odobenidae. Two subspecies are recognized based on geographic, morphological, and genetic differences: *O. r. rosmarus* in the North Atlantic and *O. r. divergens* in the North Pacific. A third proposed subspecies, the Lapteev walrus, was found to be an isolated population of the North Pacific walrus inhabiting the Lapteev Sea.

Walruses were commercially hunted from the eighteenth through to the mid-twentieth centuries for oil, meat skin, and ivory, and some subsistence hunting continues today among indigenous people. A study of DNA from walrus bones uncovered the existence of a genetically unique population of walruses that was hunted to extinction by Norse settlers during the eleventh and twelfth centuries. The evolutionary history of walruses reveals a considerable diversity with at least 20 described fossil species in 16 genera found in middle Miocene-Pleistocene (16–<1 mya) sediments along both the Pacific and Atlantic coasts.

Both males and females have tusks, which are large canine teeth, but they are generally bigger, thicker, and longer in males. Tusks grow throughout life. Walrus tusks can grow to be 3 feet (1 m) in length and in large males weigh more than 11 lb (5 kg). The tip of the tusk has an enamel coating that can be abraded and worn away by sediment when the animal is feeding. Several functions for the tusk have been proposed: to assist them in hauling out on ice, to excavate mollusks, and in male-to-male aggressive encounters. Observations of walrus behavior indicate that the tusks are mainly used by males in social interactions, and that a male will fight if another male intrudes on him during a courtship display. The strongest males with the largest tusks typically dominate social groups. Tusks are employed to a lesser extent to haul their heavy bodies along the ice. They are not used in feeding.

Walruses swim primarily moving side to side with propulsion coming from the hind limbs like phocids. Unlike phocids, the hind flippers can be rotated forward under the body, which enables them to "walk" on land like otariids.

Walruses are highly specialized feeders eating mainly bivalve mollusks (clams). They use their small foreflippers to swim along the sea bottom in a head down position, using their sensitive whiskers to locate prey.

SIZE

Males: length 12 ft (3.6 m); weight 1,940 lb–3,439 lb (880 kg–1,560 kg)
Females: length 10 ft (3 m); weight 1,278 lb–2,290 lb (580 kg–1,039 kg)

DIET

Primarily benthic invertebrates (clams, worms, snails, and shrimp)

HABITAT

Relatively shallow shelf areas

LIFE HISTORY

Sexual maturity
Males: 7–10 years
Females: 7–8 years
Gestation: 5–16 months
Reproduce: 2–3 years

STATUS

Vulnerable

ABOVE
A walrus with its tusks displayed on the coast of Wrangel Island, Russia.

OVERLEAF
Walrus colony gathered on a beach in Svalbard, Norway.

## WALRUS IVORY

The granular material filling the pulp cavity of the tusk is structurally unique and characteristic of walrus ivory. The tusks of walruses are often carved into jewelry and artwork by the native people of Greenland, North America, and Russia. International trade of ivory is regulated. In the US, Alaskan natives are permitted to hunt walruses and to sell carved ivory products in a sustainable manner, although several states have enacted broader bans on ivory that include walrus, mammoth, and fossilized ivory.

# Ringed Seal

## The most common and widely distributed Arctic seal, the ringed seal gets its name from the marked silver rings that encircle dark spots on the pelt.

Ringed seals (*Pusa hispida*) are circumpolar; they are found in all ice-covered seas in the Northern Hemisphere and they are native to the Arctic Ocean. There are five distinct geographic forms of ringed seals that occupy northern regions of Canada, Alaska, Greenland, the Baltic Sea, Lake Ladoga in Russia, Lake Saimaa in Finland, and the Sea of Okhotsk near Japan.

Ringed seals spend most of their lives with ice floes and pack ice, using their foreflipper claws to excavate breathing holes in the ice and occupy territories with breathing holes and underwater areas beneath them. Ringed seals build snow caves and use them for giving birth to and rearing their young, as well as for protection from predators. The construction of these dens is highly dependent on sufficient annual snowfall.

Ringed seals are opportunistic feeders, but adults show preference for small schooling fish. Warming of Arctic ice has affected food webs and the trophic position of ringed seals. Those seals living in the high Arctic appear more detrimentally affected compared to seals in the mid Arctic due to changes in zooplankton populations. In particular, in the high Arctic copepod populations at the base of the food chain have become dominated by fewer, less lipid-rich species, decreasing the availability of energy-rich lipids for higher trophic levels such as fish prey and ringed seal consumers. A consequence of this is a negative impact on the body condition of seals, such as decreasing blubber thickness.

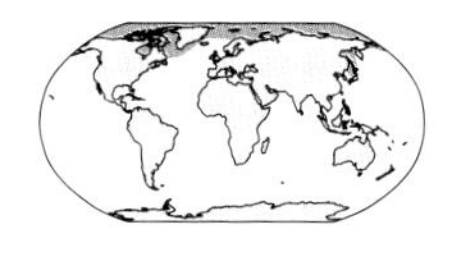

SIZE

Males & females
Length: 3 ft 7 in–5 ft (1.1 m–1.6 m)
Weight: 110 lb–198 lb (50 kg–90 kg)

DIET

Many species of fish, cephalopods, crustaceans

HABITAT

Ice-covered Arctic seas

LIFE HISTORY

Sexual maturity
Males & females: 8–10 years
Gestation: 9–11 months
Reproduce: yearly

STATUS

Least concern
Threat from habitat degradation; in the past hunted

A ringed seal showing the full rounded shape and ringed coat pattern for which they are well named.

Ringed seal emerging from a breathing hole in the Arctic. They are especially vulnerable and subject to capture in their breathing hole by their chief predator, polar bears.

Ringed seals are among the smallest phocid seals. They have exceptionally well-developed vibrissae or whiskers. A single whisker of the Baltic ringed seal contains ten times the number of nerve fibers of land mammals. Ringed seals can sense water-borne sound waves that detect changes in swimming speed and direction, which may help them find their way in the dark and through cloudy water beneath the ice.

Because ringed seals are so dependent on Arctic ice, and rarely come onto land, the loss of sea ice has detrimentally affected them in several ways. Rising sea temperatures result in the early break up of sea ice, which can cause nursing young to be separated from their mothers earlier than would have typically occurred. Insufficient snow can cause the collapse of dens, leaving ringed seals unsheltered and exposed to predators. Warmer ocean temperatures increase the spread of parasites and pathogens, which is facilitated by the migration of seals as they are forced to find more stable ice habitats.

# Amazon River Dolphin

## The Amazon river dolphin is also known as the boto or pink river dolphin. Its distinctive and varying color patterns are dependent on age and range from white/pink to totally vivid pink.

Amazon river dolphins (*Inia geoffrensis*) are probably the best-studied and most well-known of river dolphins. Two subspecies have been recognized, *I. g. geoffrensis* and *I. g. boliviensis*, although their validity is questioned.

Amazon river dolphins are the largest species of river dolphin. Male dolphins are pinker than females. The coloration is thought to be a product of scar tissue resulting from rough play or fighting. However, their final color can be influenced by their behavior, capillary placement, diet, and exposure to sunlight, with brighter pinks attracting more attention from the females during mating season. Botos are endemic to the Amazon and Orinoco drainage basins of South America, occupying rivers with varying levels of connectivity in Brazil, Colombia, Ecuador, Bolivia, and Venezuela.

Botos have bulbous heads, long snouts, and uniquely differentiated teeth. The conical, anterior teeth are used to grasp prey, heavily scaled fish, including piranhas, and bottom-dwelling crustaceans, which are crushed by the molars in the rear of the mouth before being swallowed. Amazon river dolphins possess stiff hairs (whiskers) on the beak that help them to detect prey in murky water along with echolocation. Unlike other river dolphins, the boto has a flexible neck, which allows it to move its head between trees and tangled branches. They have a varied diet, feeding on more than 50 different species of fish.

Botos are not highly social. Loose aggregations of 12–19 individuals have been observed generally at river bends or confluences, where aggregations form for mating and feeding. More typically they occur singly or in small groups of two to three. Typically, social bonds occur between mother and calf, but bachelor herds have also been observed. Male Amazon river dolphins have been seen carrying objects such as sticks, branches, and clumps of grass. Significant aggression between adult males, such as biting or striking another dolphin with the head or tail, have been associated with object carrying and were linked to increased access to females.

A model of population size based on the survival and reproduction rates suggests that the Amazon river dolphin will decline 95 percent in 50 years and may be extinct in a century. River dwelling cetaceans are particularly vulnerable to human pressures because they cannot escape them. These human threats include

SIZE

Males: length
6 ft 6 in–8 ft (2 m–2.5 m); weight 409 lb (185 kg)
Females: length 5 ft–7 ft (1.6–2.2 m); weight 330 lb (150 kg)

DIET

Bottom-dwelling fish

HABITAT

Amazon and Orinoco River drainages

LIFE HISTORY

Sexual maturity
Males & females: 5 years
Gestation: 11 months
Reproduce: 1–3 years

STATUS

Endangered
Threats from pollution, fisheries bycatch

Amazon river dolphins swim in a flooded forest in Rio Negro, Brazil.

pollution, fragmentation of waterways, hunting, and accidental entrapment in fish nets. World Wide Fund for Nature is using drones to count botos, a method which is quicker, cheaper, and more accurate than traditional monitoring. Botos are also equipped with satellite tags to monitor their migration behavior.

The Amazon river dolphin is a member of the family Iniidae and is most closely related to the La Plata river dolphin.

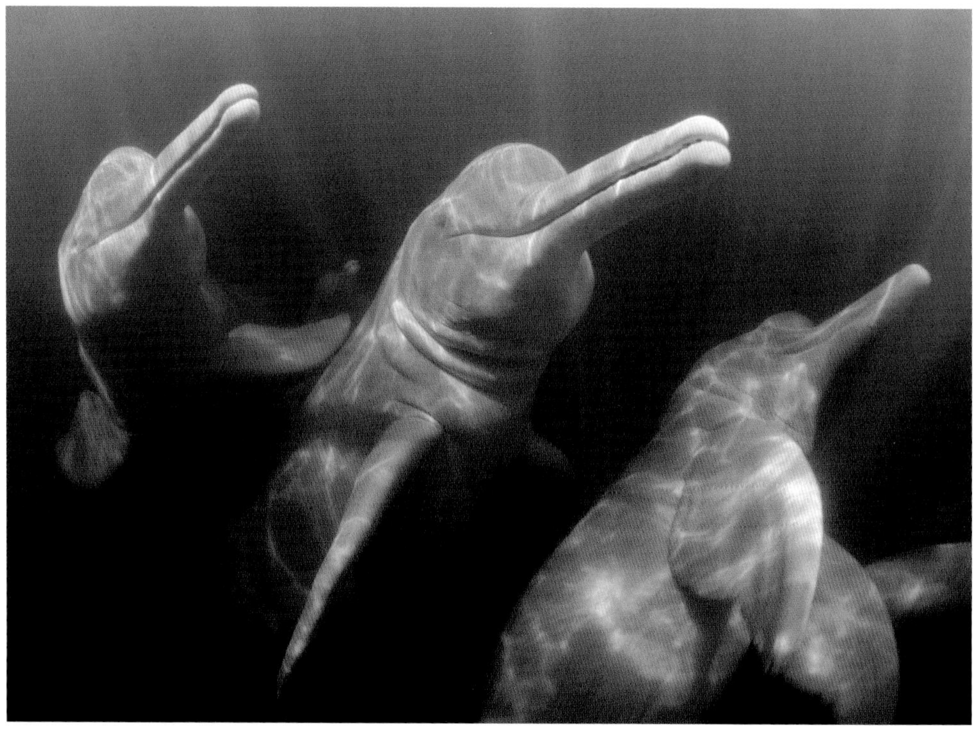

OPPOSITE TOP
Amazon river dolphin illustrating the vivid pink coloration typical of some male individuals of this species.

OPPOSITE BOTTOM
Amazon river dolphins have characteristic long, narrow beaks, tiny eyes, and short, broad flippers.

## Evolution of River Dolphins

The Amazon river dolphin is one of four genera of river dolphins. A second genus of river dolphin also lives in South America, the La Plata river dolphin or franciscana, *Pontoporia blainvillei*. The Yangtze river dolphin, *Lipotes vexillifer* (see page 194), occupies the Yangtze River in China; the Indian river dolphin, *Platanista gangetica*, lives in the Ganges River and *P. minor* occupies the Indus River of India. Despite the fact that river dolphins share a number of similarities and traditionally were grouped together, they are not closely related.

River dolphins share long, thin rostrums, reduced eyes, numerous teeth in upper and lower jaws, and flexible necks. These features have been considered adaptations for living in riverine habitats. For example, it was suggested that since river dolphins occupy muddy rivers with low visibility they relied on echolocation rather than vision. Unlike the stiff neck vertebrae of most toothed whales, the flexible necks of river dolphins were suggested to help them navigate through flooded, tangled forests. Study of these and other features in an evolutionary framework, however, reveals that many of them are not adaptations for freshwater environments but were already present in their marine ancestors prior to the entry into fresh water. This fits with the current hypothesis of the origin of river dolphins. According to this hypothesis, the ancestors of living river dolphins were coastal marine species that were stranded in far-flung river systems in South America, India, and China as sea level dropped during the Cenozoic. This hypothesis suggests that tolerance for fresh water evolved at least three times in river dolphins (see dashed lines below).

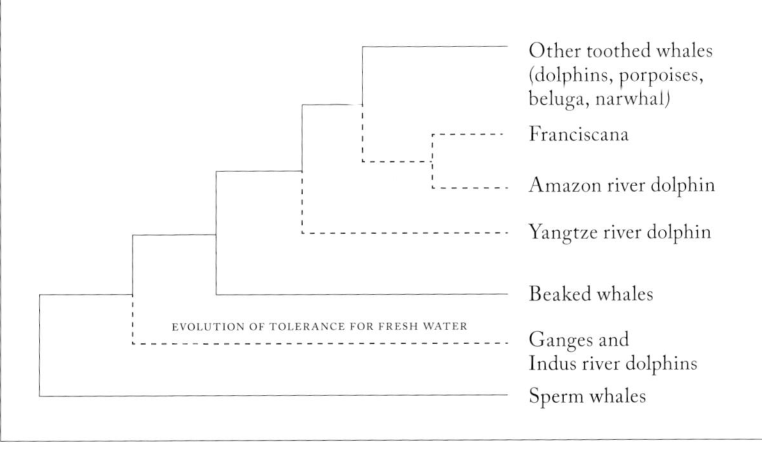

# Ribbon Seal

## One of the most striking, easily recognizable species, the ribbon seal is named for the coloration of light-colored bands or "ribbons" that encircle the neck, foreflippers, and hips on its dark coat.

Ribbon seals (*Histriophoca fasciata*) occupy the North Pacific Ocean and adjacent southern part of the Arctic Ocean. They occur most commonly in the Sea of Okhotsk and the Bering Sea. They are found in US waters off the coast of Alaska, the Bering Sea, and in the Chukchi and western Beaufort Seas.

Ribbon seals are solitary, spending most of their time in the open ocean and forming loose groups in pack ice to give birth, nurse pups, and molt. Their movement on ice differs from the typical caterpillar-like undulations of other seals. They alternate foreflipper strokes to pull themselves forward while moving their head and hips in a side-to-side motion. Ribbon seals are known to eat a variety of fish, squid, and octopus. They have a unique anatomical feature of the respiratory system, a large inflatable air sac connected to their trachea. It is larger in males than in females, and its function is unknown, although it is thought it could be used to produce underwater vocalizations or it may provide extra buoyancy when inflated, making it easier for the seals to float and rest in water.

Studies of ice-associated seals including ribbon, spotted, and some harbor seals in the Arctic have revealed declines in the body condition of pups, likely due to poor foraging conditions for their mothers during pregnancy and while nursing.

Ribbon seals may be particularly sensitive to loss of sea ice since they are more adapted for deep diving. As the ice edge of the Bering Sea recedes northward into shallower water, ribbon seal mothers and pups may find themselves some distance from foraging grounds. Measurements of body condition showed a trend toward reduced body mass and skinnier seals. Warm conditions are predicted to continue with detrimental impacts on ribbon seals and other ice-associated seals.

Large commercial harvests of ribbon seals were carried out from the 1930s to the mid-1990s, and small localized harvests continue today. Subsistence hunting by Alaska natives occurs at low levels in the US.

Ribbon seals are members of the family Phocidae and their closest relatives are harp seals.

SIZE

Males & females: length 6 ft (1.8 m); weight 198 lb 6 oz–326 lb (90 kg–148 kg)
At birth: length 2 ft 7 in (0.8 m); weight 23 lb (10.5 kg)

DIET

Fish and invertebrates

HABITAT

Arctic and subarctic

LIFE HISTORY

Sexual maturity
Males: 3–5 years
Females: 2–4 years
Gestation: 11 months
Reproduce: yearly

STATUS

Least concern
Threat from fishing gear entanglement; in the past commercially hunted

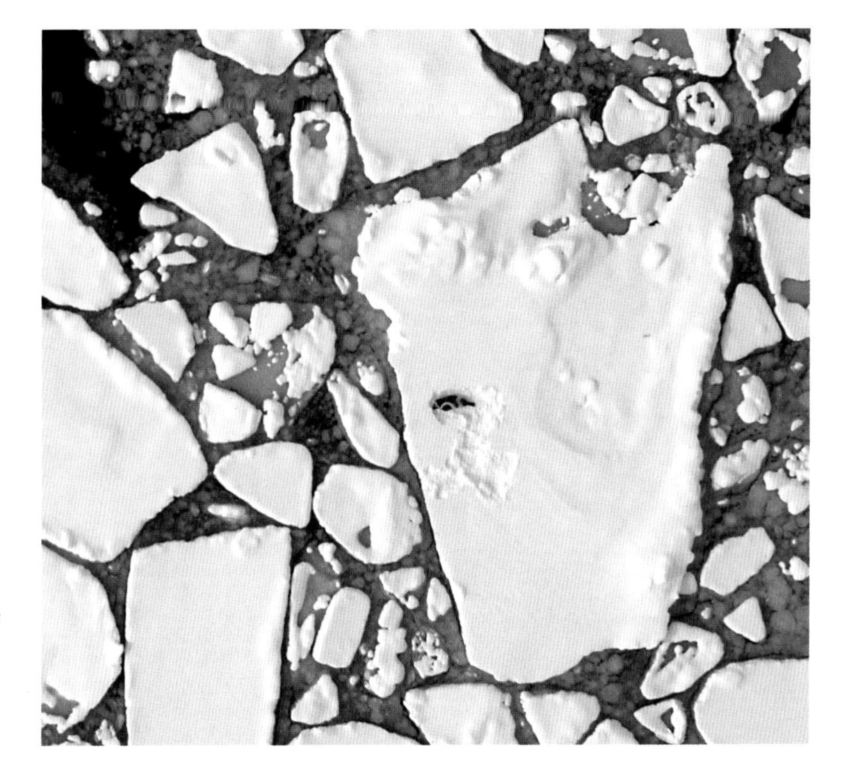

RIGHT
The bold banded coloration pattern of adult male ribbon seals makes them very conspicuous on the ice.

BELOW
Ribbon seals have a distinctive pattern of movement in which they pull themselves across the ice using alternate strokes of the foreflippers.

# BAIKAL SEAL

## The Baikal seal is the sole seal species that lives primarily in fresh water and is found only in Lake Baikal, the world's largest lake in Siberia.

Baikal seals (*Pusa sibirica*) are among the smallest seals. They are top predators and the only endemic mammal of Lake Baikal. Males and females are similar in size. Baikal seals are dark silver-gray on the back and light gray on the underside. Relative to body size, they have the largest eyes of any pinniped, and their huge eyes are thought to be an adaptation for hunting fish in deep water.

In addition to fish, Baikal seals appear to devour small crustaceans known as amphipods. When feeding, the seals use their comb-like postcanine teeth to expel water while retaining prey. Feeding on this extremely available prey could explain why Baikal seals are so abundant. The foreflippers and claws of this species are large and more strongly developed than other seals, suggesting they are adaptations for making and maintaining breathing holes and grasping prey. Scientists have recently discovered that the teeth of the Baikal seal provide data on the effects of pollution, nuclear testing, and climate change on Lake Baikal. Toxic metals such as uranium, mercury, cadmium, and zinc trapped in tooth layers date back to before industrialization of the area, and provide clues to the past environment of the lake.

Baikal seals are solitary for most of the year. They haul out on islands and shorelines during ice-free periods in the summer. In the autumn when ice forms they live on the ice. Mating takes place in the water. Similar to ringed seals, females give birth in snow-covered lairs excavated in the lake for protection from predators. Like the ringed and ribbon seals, pups are born on the ice, with a white woolly coat that is eventually replaced by the adult silver-gray coat.

Baikal seals were hunted for their skins, meat, and blubber by indigenous peoples as well as being victims of commercial hunting that continues today.

Baikal seals are members of the family Phocidae most closely related to ringed seals. When first named in 1788 it was thought that Baikal seals were just a form of common seal and not a distinct species.

SIZE

Males: length 4 ft–4 ft 7 in (1.2 m–1.4 m)
Females: length 4 ft (1.2 m)
Both sexes: weight 110 lb–198 lb (50 kg–90 kg)
At birth: length 2 ft (60 cm); weight 9 lb (4 kg)

DIET

Sculpins, oilfish

HABITAT

Lake Baikal

LIFE HISTORY

Sexual maturity
Males: 7–10 years
Females: 3–7 years
Gestation: 9 months
Reproduce: yearly

STATUS

Least concern
Threats from pollution, commercial and subsistence hunting, fisheries bycatch

RIGHT
Large eyes and claws are distinctive features of the Baikal seal.

BELOW
Baikal seals typically haul out in groups.

OPPOSITE TOP
A group of Baikal seals hauled out on a crowded rookery in the summer.

OPPOSITE BOTTOM
Baikal seal showing the light gray belly and dark gray back.

## HOW DID BAIKAL SEALS GET TO A LAKE IN THE MIDDLE OF RUSSIA?

The question of how the Baikal seal came to occupy a land-locked lake in the first place is still debated. Two hypotheses have been proposed, either a Paratethyan origin or an Arctic origin. A Paratethyan origin suggests that living *Pusa* seals (Baikal, Caspian, and ringed seals) descended from fossil ancestors that occupied Paratethys, an inland sea that covered most of southeastern Europe and southwestern Asia during the Oligocene-Miocene (35–20 mya). From this region seals invaded the Caspian Sea and dispersed further east to Lake Baikal. The alternate hypothesis posits that Baikal seals have not descended from Paratethyan seals that migrated south to north, but that three *Pusa* seals are of Arctic ancestry and the seals got to Lake Baikal from the north. This hypothesis is favored and further support comes from the fact that large ice-dammed lakes occupied central Siberia about 300,000 years ago. These lakes apparently had connections to the Arctic Ocean, via the Yenisey-Angara River systems. This hypothesis better explains the ecology and life history of the Baikal seal and it is also supported by plausible dispersal routes that would have facilitated entry of the Baikal seals into Lake Baikal.

# 4
# BEHAVIOR

Highlighted in this chapter
are the remarkable array of
behaviors displayed by sea mammals
that have been shaped primarily
by where they live, how they move,
where they feed, and what they eat.

Feeding strategies among sea mammals are influenced by habitat, prey availability, and social learning. Highly intelligent common bottlenose dolphins employ several cooperative feeding strategies. Indo-Pacific dolphins of Shark Bay, Australia, have developed a unique foraging strategy that involves using sponges to protect their beaks as they flush fish and other prey from crevices. Their social structure varies from stable to weak depending on the degree of environmental stability. They are one of the few mammal species in which males form alliances and cooperate with other males to allow for easier access to females for mating.

Like elephant seals, Weddell seals are champion divers and are capable of diving more than one mile on a single breath hold. Sperm whales are also expert divers and feed on squid at depth; they make click sounds to locate prey and to talk to each other. Both Weddell seals and sperm whales are able to make long dives due to a variety of respiratory and circulatory adaptations, including storing oxygen in their muscles and blood rather than in their lungs as humans do.

Among the most acrobatic of toothed whales, spinner dolphins likely communicate their position to other dolphins, play,

PAGE 124
A dolphin surfing waves in Australia. The power of waves optimizes swimming, allowing dolphins to swim with less effort.

and remove parasites through aerial spinning behaviors. Belugas are high-latitude animals and molt in the summer in warmer waters to maintain healthy skin. They live in complex social systems involving interactions between kin and non-kin.

The fastest cetaceans are Dall's porpoises, which leave a "rooster tail" of spray when traveling at high speed. This species also employs a risky reproductive strategy known as mate guarding to ensure that males protect their mates from potential rivals. The West Indian manatee, like other manatees, possesses special body hairs that allow them to sense water movements, such as those of prey in an underwater world where vision is limited.

Shark Bay groups of male dugongs gather in leks to participate in competitive activities and displays to ward away other males and attract females during mating.

One of the few pinnipeds living in subtropical and tropical environments, Galapagos fur seals cool their bodies by employing a behavioral strategy known as "sand flipping" in addition to seeking out rock crevices to escape the heat.

# Common Bottlenose Dolphin

## The common bottlenose dolphin is the most well known of any sea mammal. Some dolphins exhibit a unique feeding strategy using mud rings to herd fish.

Common bottlenose dolphins (*Tursiops truncatus*) have a generalized appearance with a medium-sized, robust body, a curved dorsal fin, and a sharp demarcation between the melon and the short rostrum. They are light gray to black on the back with a light belly. A distinct cape may be visible unless the dolphin is very darkly colored.

The taxonomic status of the genus *Tursiops* is unclear primarily because of geographic and morphological variability within the genus. In addition to *T. t. truncatus*, the common bottlenose dolphin, two other subspecies are recognized: *T. t. ponticus*, the Black Sea bottlenose dolphin restricted to the Black Sea, and *T. t. gephyreus*, Lahille's bottlenose dolphin occupying the South Atlantic.

Bottlenose dolphins have a worldwide distribution in both coastal and offshore cool temperate to tropical waters. It is the coastal habitat of some dolphins that brings them into close contact with humans, making them the most widely recognized small cetacean. They were among the first marine mammals to be kept in captivity and they are often the charismatic stars of aquarium shows.

Bottlenose dolphins are among the fastest marine mammals, swimming up to 18 mph (29 kmh) in short bursts with an average speed of 5 mph (8 kmh) when traveling. They are often seen body surfing in the waves, or following the wake of boats—an energy saving strategy. Dolphins also conserve energy by leaping out of the water, as much as 20 feet (6 m) high, which removes them from the turbulence and high drag of waves at the water surface.

In addition to producing high-frequency echolocation sounds for navigation and to locate food, bottlenose dolphins also produce whistles. The distinctiveness of an individual dolphin's whistle, known as a signature whistle, function like names to broadcast the identity of the animal producing the whistle, and possibly to communicate other information, such as their state of arousal or fear, to group members.

The dolphin brain like that of other toothed whales is relatively large and similar in size to a gorilla's brain. Only the human brain is proportionally larger. The complex cognitive abilities of dolphins are well known. The high encephalization quotient of dolphins, a measure of brain size in relation to body size, is partly explained by their complex social structure and behavior.

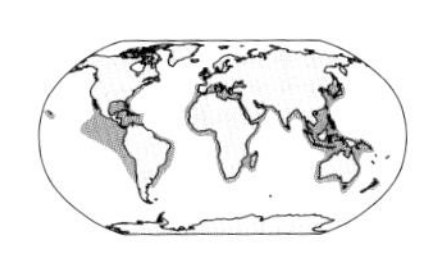

SIZE

Males & females
Length: 6 ft–13 ft (1.8 m–4 m) males slightly larger
Weight: up to 1,433 lb (650 kg)
At birth: length up to 4 ft (1.2 m); weight 22 lb–44 lb (10 kg–20 kg)

DIET

Mostly fish

HABITAT

Coastal and continental shelf waters

LIFE HISTORY

Sexual maturity
Males: 9–13 years
Females: 5–13 years
Gestation: 12 months
Reproduce: 3–6 years

STATUS

Least concern
Threat from habitat degradation

A common bottlenose dolphin leaps out of the water; this action is an energy conservation measure.

Common Bottlenose Dolphin

# Indo-Pacific Bottlenose Dolphin

## The Indo-Pacific bottlenose dolphins living in Shark Bay have a unique foraging strategy that involves using a sponge as protection for their rostrum.

Recognition of Indo-Pacific bottlenose dolphins (*Tursiops aduncus*) as a distinct species most closely related to bottlenose dolphins is based on genetics, osteology, and external anatomy. Until 1998 all bottlenose dolphins were classified as common bottlenose dolphins, *T. truncatus*. Indo-Pacific bottlenose dolphins differ from common bottlenose dolphins in being smaller and possessing more slender teeth, their beak is narrower and longer, and they are lighter in color. The belly is generally off-white to gray, grading to a darker gray on the sides and the back. The dark gray cape is generally more distinct and extends to the tail stock.

Indo-Pacific bottlenose dolphins occupy warm temperate to tropical nearshore waters ranging from South Africa in the west to eastern Australia and the Solomon Islands/New Caledonia in the east. They are one of the most frequently observed cetacean species on continental shelves, and around islands of the Indian and western Pacific Oceans. They are found in a variety of habitats ranging from rocky and coral reef areas to sandy bottoms and seagrass beds. They feed on a variety of schooling and reef fish as well as cephalopods (e.g., squid).

Like most dolphins they are highly social, commonly occurring in groups of 20, but they are sometimes found in groups of 100 or more. Indo-Pacific bottlenose dolphins have a fission-fusion social structure and they are one of the few mammal species in which males cooperate with other males. Males form alliances or coalitions with one to three other males and these males herd females for mating. Males typically are more heavily scarred than females and may show rake marks on their dorsal fins, the result of aggressive encounters with other males. Breeding females also form groups, that when large make them difficult to defend. Juveniles form behaviorally specific associations, engaging in play that allows them to learn social skills.

Indo-Pacific bottlenose dolphins produce numerous types of vocalizations in addition to echolocation clicks, including diverse types of whistles related to the type of behavior that they are engaged in. The majority of whistle repertoires are unique and likely primarily driven by their complex social organization.

The major threat to this species is habitat degradation due to coastal development that places them near human activities. They are also hunted in Sri Lanka and the Solomon Islands and are victims of incidental catches in the Taiwanese driftnet fishery operations in Indonesia.

SIZE

Males & females
Length: 7 ft (2.1 m)
Weight: 507 lb (230 kg)

DIET

Schooling and reef fish, cephalopods

HABITAT

Warm-temperate to tropical coastal waters

LIFE HISTORY

Sexual maturity may occur at lengths of 8 ft (2.5 m)
Gestation: 12 months
Reproduce: 4–6 years

STATUS

Near threatened
Threats from incidental bycatch, habitat degradation

Indo-Pacific bottlenose dolphins located in the Red Sea, Egypt, show the distinctive coloration pattern of a dark gray back and lighter pale gray to white on the sides and belly.

Indo-Pacific bottlenose dolphin showing the distinctive coloration pattern.

## DOLPHIN TOOL USE

Over the last 30 years, studies of tool use in Shark Bay Indo-Pacific bottlenose dolphins has shown that sponging dolphins associate with other sponging dolphins and their alliance is based on where they choose to forage. Sponging involves dolphins carrying conical sponges as protective "gloves" on their rostra when foraging. This behavior may help spongers find food in deep-water channels, since when they find a hidden fish, they drop the sponge and grab the fish with their teeth. This strategy is favored in the deep-water channels of Shark Bay, which are known to be of lower productivity in terms of quality and quantity of fish. This behavior appears to be primarily culturally transmitted from mother to female offspring. Although mothers show both males and female calves how to use sponges, female calves are almost exclusively the only ones to apply this knowledge. It is less common among males who feed in sand flats and focus on male-male alliances that are largely incompatible with the time-consuming, solitary, and difficult sponging activity. Recent testing of the link between sponging and male behavior revealed that male spongers associate significantly more often with other male spongers irrespective of their degree of relatedness. Unexpectedly, the time spent socializing was equal among male spongers and non-spongers. Spongers did not seem to have a competitive advantage over non-spongers, as females in both groups produced the same number of calves.

Besides sponging, researchers in several areas of Shark Bay have also observed another form of tool use, in which individual Indo-Pacific bottlenose dolphins lift large, abandoned conch shells out of the water. The dolphin's rostrum was inserted into the broad shell aperture and a trapped fish taking refuge in the shell was eaten. Feeding was the predominant activity recorded for individuals when conching (or shelling) was observed. Research showed that unlike sponging, conching behavior is learned from peers. Although the conching dolphins were not closely related, analysis showed that they belonged to the same social networks. The more time that two individuals spent together the more likely they were to copy shelling behavior from one another.

# WEDDELL SEAL

## Weddell seals are the world's most southern breeding mammal. They are also extraordinary divers, capable of dives over 1,960 feet (600 m) that can last up to 80 minutes on a single breath hold.

Weddell seals (*Leptonychotes weddellii*) are among the largest and most abundant marine mammal in Antarctica. Weddell seals occupy ice attached to land, known as fast ice, and offshore seas. They exhibit a countershaded coloration pattern with a dark gray back and lighter gray or silver belly. Adult females are slightly larger than adult males.

The diving physiology of Weddell seals has been extensively studied. In the 1960s, pioneering studies began on Weddell seals in the wild, and deployed time depth recorders (TDRs). Most of the long exploratory Weddell dives are to locate breathing holes in the ice. Weddell seals use their teeth, especially procumbent canines, to create air holes. These dives are done anaerobically and require time at the surface to replenish oxygen stores. Numerous adaptations of their circulatory and respiratory systems facilitate deep diving, including flexible ribs that allow the lungs to collapse, oxygen that is stored in muscles and blood rather than the lungs, reduced oxygen consumption, and a slowed heart rate. A second major type of dive, primarily to locate food, is shorter (8–14 minutes in duration), shallower (up to 1,300 feet [400 m]), and aerobic, and does not require adjustments to the respiratory and circulatory systems. The diving ability of pups increases with age. Study of the genome reveals the regulation of genes involved in the diving response, and expression differences between adults and pups may explain the time required for the development of diving capacity that results in an increase in dive duration and depth.

More recently satellite relay tags (known as DSA tags) provide finer scale information on body temperature and diving behavior of the seal. From these measurements foraging activity and energy expenditures of seals can be estimated. In one study a high correlation was found between vertical speed and the number of prey capture attempts, indicating that even with low-resolution dive data, often the case in remote areas, an assumption can be made with confidence about hunting segments of dives. Another study showed that there was variability in the diving behavior of individuals. Some seals showed little dispersal and foraged within a small core area, while other individuals ranged further into deeper waters with a high ice concentration. This has conservation implications, suggesting that protecting continental shelf areas would benefit Weddell seals, a finding that is confirmed by other data.

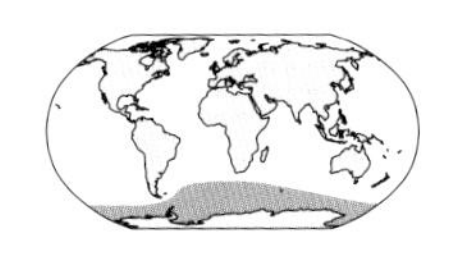

SIZE

Males: length up to 10 ft (m)
Females: length up to 11 ft (3.6 m)
Both sexes: weight 882 lb–1,323 lb (400 kg–600 kg)

DIET

Antarctic toothfish, squid

HABITAT

Fast ice

LIFE HISTORY

Sexual maturity
Males: 7–8 years
Females: 3–6 years
Gestation: 11 months
Reproduce: yearly

STATUS

Least concern
Threat from climate change

A Weddell seal shows off its distinctive coloration—a dark gray back with a light gray to off-white belly. It rests on an ice floe in Antarctica.

Weddell seals are sentinels of sea ice changes, allowing researchers to collect winter data that had previously been unattainable due to harsh conditions. To better understand continental ice shelves, Weddell seals were outfitted with satellite tags providing readings on conductivity, temperature, salinity, and the depth of the ocean. Scientists confirmed seasonal changes in sea ice, such as the presence of warm and low-salinity water, which had positive effects on the seals' foraging behavior.

The diet of Weddell seals varies regionally and seasonally. The principal prey is fish, mostly Antarctic silverfish as well as Antarctic toothfish, cephalopods, and crustaceans. Research has shown that they are stealthy hunters and sneak up on fish at close range. They also use a method of disturbing fish from ice cracks by blowing bubbles into them and preying on the fish that emerge.

OPPOSITE TOP
A Weddell seal mother plays with her pup. These seals haul out on fast ice to rest, molt, and pup.

OPPOSITE BOTTOM
A female Weddell seal swims under sea ice at Signy Island, Antarctica. This species is known for its exceptional diving abilities. They are capable of diving for nearly 1.5 hours at depths of 1,968 feet (600 meters).

Weddell seals are unique among Antarctic seals in forming colonies with strong site fidelity. Adult females during breeding form loose aggregations along cracks in the ice that sometimes exceed hundreds of individuals. Adult males maintain underwater territories and control females' access to breathing holes. The potential for polygyny is high since the ephemeral structure of fast ice forces females to gather together. The level of polygyny, however, is lower than for other species and the harems are usually small, up to five females per male. Individuals form mother-pup pairs.

Males guard their territories using loud vocalizations that include trills, chirps, and whistles. Studies have shown that some of these sounds (perhaps as many as one-fifth of their underwater vocalizations) are produced at ultrasonic frequencies, above the range of human hearing. Although it is not known why they produce high-frequency sounds, it has been suggested that these high-pitched sounds may help them navigate under the sea ice, especially at night. They also may produce ultrasonic vocalizations for communication, thereby allowing them to stand out from other low-frequency noise. No other pinnipeds are known to employ sounds at ultrasonic frequencies.

In one of the first attempts to assess Weddell seals' global population, high-resolution satellite imagery and citizen scientists counted images of Weddell seals in the Antarctic, and revealed a lower number of seals than expected. The study also found that the seals inhabited a very small portion of fast ice, so called because it is connected to the coastline; this is a critical habitat because Weddell seals give birth there. Seal colonies were found associated with habitat preferences, such as deep water and prey availability, and this information can be used to help preserve specific Weddell seal habitats in the future.

Weddell seals are in the pinniped family Phocidae, subfamily Monachinae (Southern Hemisphere seals), and they are most closely related to leopard seals.

# Sperm Whale

## The sperm whale is the largest toothed whale. The name comes from the large spermaceti organ in the head, once believed to carry sperm, but now known to contain fluids that are used in echolocation.

Sperm whales (*Physeter macrocephalus*) are perhaps best known from Herman Melville's classic whaling novel *Moby-Dick,* although they are not ferocious beasts capable of sinking ships and killing sailors. Sperm whales possess teeth only on the lower jaw, which are used to suction feed deep-sea squid and fish.

Sperm whales are predominantly black to brownish gray in color. They differ from other whales in having a square-shaped head that accounts for about one-third of the body length. Sperm whales display the greatest degree of sexual dimorphism of any known whale. The single blowhole is asymmetrically positioned on the left. They possess two to ten short throat grooves that facilitate suction feeding. The dorsal fin and paddle-shaped flippers are small.

Sperm whales are distributed in all the world's oceans. They tend to occupy continental slopes and deep oceanic waters. Sperm whales are the deepest diving vertebrates, staying submerged for well over two hours (138 minutes) and diving to depths of more than 1.8 miles (3,000 m) on a single breath! Study of the diving behavior of sperm whales from different localities reveals that most commonly during foraging they dive to 1,312 feet (400 m) for 35–40 minutes to exploit food patches using echolocation for locating prey.

Sperm whales are very social animals. Females and their young are found in the tropics and subtropics traveling in social groups that average about a dozen whales, known as units. These long-term matrilineal families are similar to the social organization of African elephants and killer whales. Regional differences were found in the matrilineally based social structure, which may relate to prey characteristics affecting group size similar to killer whale ecotypes (see page 92) and diets. Alternatively, these differences could be the population effects of whaling. For the latter the importance of kinship was diminished in more heavily hunted populations, which is also the case for African elephants.

Sperm whales are polygynous and young males leave their family groups between the ages of 4 and 21 to form fluid bachelor herds that rove in higher latitudes between breeding and feeding grounds, as well as among females when breeding. Study has found evidence that male sperm whales can develop long-lasting bonds, forming friendships with other males that can last for at least five years. These male bonds may enhance their ability to find food by sharing

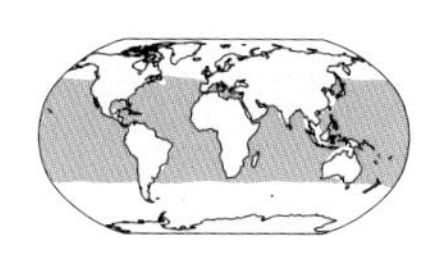

SIZE

Males: length up to 63 ft (19.2 m); weight 88 tons (80 tonnes)
Females: length up to 41 ft (12.5 m); weight 27.6 tons (25 tonnes)

DIET

Squid, deep sea fish

HABITAT

Deep oceans

LIFE HISTORY

Sexual maturity
Males: 19 years
Females: 9 years
Gestation: 15–16 months
Reproduce: 5–7 years

STATUS

Vulnerable
Threat in the past from commercial hunting

A sperm whale ascends to the surface, it is distinguished by its large size and blunt head.

information about prey. The roving strategy of adult males may be assisted by long-distance vocal communication between traveling males and female social groups.

Groups of females and young occupy tropical and subtropical waters year round, whereas male sperm whales form fluid bachelor herds and migrate over long distances and may be found at higher latitudes of both poles.

Sperm whales produce a variety of sounds known as clicks that are used in different contexts. The unique low-frequency click series that male individuals produce, known as codas (with a Morse code-like pattern), might help recognition of individuals between and outside groups. Codas appear to be culturally acquired from within family units similar to killer whale dialects. Clicks are also used in echolocation helping them to locate food. They also produce pops or bangs: loud, low-frequency sounds that likely debilitate or stun prey.

Sperm whales display an unusual sleep pattern. They sleep for about 10–15 minutes at a time by diving down, turning around, and napping in a vertical position while they slowly float back up to the surface.

Sperm whales were the primary whale hunted by Yankee whalers in the northeastern US coast. Oil was the primary target but ambergris was even more valuable. From the mid-1940s until 1980 a second phase of sperm whale hunting involved both pelagic and land-based whaling operations. As a result, sperm whales were nearly driven to extinction. Today, sperm whales are threatened by vessel strikes, fishing gear entanglements, pollution, and climate change.

Sperm whales are early diverging toothed whales comprised of a single species in the family Physeteridae. They are most closely related to pygmy and dwarf sperm whales.

## WHAT IS AMBERGRIS?

Like cows and other ruminants, a sperm whale has a four-chambered stomach. Food is digested as it passes from one stomach to another. After numerous bouts of feeding, the stomach chambers begin to fill with undigested squid beaks that coalesce to form a large, dense glittery mass, known as ambergris. The mass is expelled from the sperm whale's mouth every few days, and washed up on to shores. Some of the earliest references to ambergris date back to 700 BCE, and fossilized evidence of the substance dates back nearly 2 million years. Arabians used trained camels to find it. The Chinese thought ambergris was dragon spit that had fallen into the ocean, and the Ming dynasty reportedly imported it from Sri Lanka and Africa. It was not until the early eighteenth century, during large-scale whaling, when the true origins of ambergris from squid beaks eaten by sperm whales were discovered. During the twentieth century, ambergris was much in demand for its use as a fixative in the making of perfume. Since then scientists developed a synthetic version and today most perfumes rely on the laboratory produced alternative.

Recent study of the DNA of ambergris reveals not just the whale's genetics but also the whale gut microbiome and potentially the DNA of their prey. This has the potential to yield significant insights into sperm whale evolution and ecology.

OVERLEAF
A pod of sperm whales made up of calves and juveniles dive in the Indian Ocean.

# SPINNER DOLPHIN

## Spinner dolphins, so named for their acrobatic behavior, repeatedly leap and spin completely out of the water.

At least five geographic forms (subspecies) of spinner dolphin (*Stenella longirostris*) are recognized. Some populations are common offshore, especially in the Eastern Tropical Pacific, whereas other populations rest in the shallow coastal waters and during the day stay in bays or coral atolls. Spinner dolphins appear to be striped and have a tripartite color pattern, with a dark gray back, a white belly, and light gray sides. They have a long beak and are highly sexually dimorphic.

Spinner dolphins are the most common small dolphin, occurring in all tropical and most subtropical waters around the world. They are generally found in groups of 100 but they sometimes gather in groups of 1,000 or more. They often congregate near divergence zones at current margins and at current ridges where there is a high concentration of small fish—their primary prey. They have been shown to cooperatively herd fish.

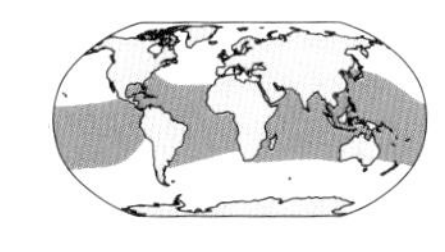

SIZE

Males: length 7 ft 6 in (2.3 m); weight 51 lb–176 lb (23 kg–80 kg)
Females: length 6 ft 6 in (2 m); weight 51 lb–176 lb (23–80 kg)

DIET

Small fish, squid

HABITAT

Oceanic tropical and subtropical waters

LIFE HISTORY

Sexual maturity
Males: 7–10 years
Females: 4–7 years
Gestation: 10 months
Reproduce: 3 years

STATUS

Least concern
Threat from fisheries bycatch

A large pod of Central American spinner dolphins swim near the surface in waters off the coast of Costa Rica.

A spinner dolphin leaps out of the water. The repeated spins provide communication among pod members and also dislodges parasites or cleaner fish.

## HOW DO DOLPHINS SPIN AND WHY?

Spinner dolphins usually perform as many as eight spins in a row, each spin tending to be made with less energy and ending with a side slap. The power of the spin comes from acceleration under the water, the rotation of the tail, and use of the flippers as wings as the dolphin breaks the surface. Underwater they generate one to two spins and airborne they can make up to seven revolutions in as little as a second. The spin can be seen from a great distance. All ages and sexes perform this behavior. But why do dolphins spin? Several explanations for this spinning behavior have been proposed, including communication to other dolphins in a group since once a dolphin starts to spin others will join in. Spinning may also dislodge parasites or remoras (a type of suckerfish) from the skin.

Spinner dolphins are highly social. Their social structure varies among the different forms of spinner dolphins. Some populations have fluid arrangements whereas others live in stable pods for many years. Among the best studied behaviorally are Hawaiian spinner dolphins. They exploit sheltered bays to socialize and rest in during the day following a night of cooperative feeding in open water. This pattern of rest and feeding allows spinner dolphins to maximize their foraging efficiency while minimizing the risk of predation during rest periods. However, long-term studies in Hawaii detected a decline in abundance since the 1990s, possibly the result of increased tourism in their resting areas. In an attempt to protect spinner dolphins from human harassment and to preserve their resting habitat, US regulators in 2021 banned swimming with the dolphins close to shore.

Spinner dolphins at more remote atolls, such as Midway Atoll and other atolls in the Northwestern Hawaiian Islands, have a different social structure with stable groups that have long-term associations. These atolls are geographically isolated and have limited areas of suitable habitat, making it energetically more beneficial to remain "at home" rather than travel to other locations.

Spinner dolphins, like pantropical spotted and common dolphins (see page 192) from the Eastern Tropical Pacific, were severely depleted during the 1960s and 1970s, victims of tuna purse seine fishing. Changes in net design and dolphin release procedures have dramatically reduced dolphin mortality caused by fishing.

Spinner dolphins are in the toothed whale family Delphinidae and they are most closely related to common dolphins, as well as to short-beaked and long-beaked dolphins.

OPPOSITE TOP
A spinner dolphin showing the tripartite color pattern.

OPPOSITE BOTTOM
School of spinner dolphins swimming at sea in Marsa Alam, Egypt. They live in large groups of a few dozen to a thousand or more in tropical and subtropical waters around the world.

# Beluga Whale

## Belugas, also known as white whales, live in Arctic waters most of the year, migrating in summer to river estuaries to molt, shedding their outer layer of skin.

The most distinctive feature of belugas is the pure white color of adults. The scientific name *Delphinapterus leucas* means "the white whale without a wing," which refers to the fact that they lack a dorsal fin, which acts as a stabilizer to control rolling during swimming. Belugas instead possess fat pads parallel to dorsal ridges along the back that are tensed by abdominal musculature enhancing the animal's ability to control heading and limit roll. Belugas are adapted to life in cold water and they have a blubber layer that is up to 5.9 inches (15 cm) thick. Their head, tail, and flippers are relatively small.

Belugas are found in cold waters of the Arctic and sub-Arctic. When sea ice recedes in spring most populations move to their summering grounds, often forming concentrations in estuaries, and sometimes rivers.

Most whales shed their skin continuously although belugas do not. Their annual molt cycle was thought to be unique to belugas, but it appears that molting may be common among all high-latitude whales, helping them to maintain healthy skin. Scientists suggest that the fresh water softens the animal's skin and they have been seen scrubbing their bodies on stony seabeds in the Canadian Arctic to stimulate skin regrowth. Instead of relying on tooth samples from dead whales to ascertain the age of belugas, a recent study shows that small skin samples can be used to determine both the age and sex of a beluga, which will enable scientists to better understand growth, reproduction, and survival rates, important in assessing a species at risk of extinction.

Belugas are highly gregarious and are often found in groups of up to 15 individuals, but groups of thousands of individuals are sometimes sighted. Groups are often segregated by age and sex, and from brief alliances to multi-year affiliations. Field observations and genetic studies revealed that belugas live in complex societies regularly interacting with close paternal kin. They also frequently associate with more distantly related kin and non-kin.

Belugas are known to be among the most vocal whales. In addition to echolocation signals used to navigate and hunt prey, they produce three types of social communication sounds: whistles, calls, and combinations of these types. The beluga and its close relative the narwhal are members of the family Monodontidae.

SIZE

Males: length 18 ft (5.5 m); weight 1.65 tons (1.58 tonnes)
Females: length 14 ft (4.3 m); weight 1.5 tons (1.36 tonnes)

DIET

Fish (salmon, herring), mollusks, crustaceans

HABITAT

Arctic and subarctic waters, seasonally in shallow coastal waters

LIFE HISTORY

Sexual maturity
Males: 8 years
Females: 5 years
Gestation: 12–15 months
Reproduce: yearly

STATUS

Least concern
Critically endangered (Cook Inlet population)
Threat in the past from commercial hunting

A beluga whale swims in the White Sea of Russia. The fat pads on the sides of the body are thought to act as vertical stabilizers that enhance the animal's balance while swimming.

# DALL'S PORPOISE

## Dall's porpoises are the fastest swimming marine mammal, reaching speeds of 35 miles per hour over short distances.

Dall's porpoise (*Phocoenoides dalli*) is one of the most common North Pacific porpoises. They have a striking coloration pattern, a black body with a white patch on the belly and lower flanks. The black and white coloration may also protect them from their major predator, killer whales. From a distance, hunting orcas might mistake Dall's porpoise for another killer whale, giving the porpoise a chance to escape in the confusion. They are the largest porpoise, and males and females have a similar stocky body with the males slightly larger and more robust. Like all porpoises, Dall's porpoise has little or no beak. Another feature of this species seen in other porpoises is their small, spade-shaped teeth. Dall's porpoises produce short, high-frequency echolocation clicks.

Two distinct subspecies of Dall's porpoise are known that were originally described as separate species, but are now known to be populational differences (subspecies) between True's porpoise, *P. d. truei*, which occurs off the Pacific coast of northern Japan, with Dall's porpoise, *P. d. dalli*, which ranges across the North Pacific from northern Japan to the Bering Sea and into California. There is evidence of hybridization between Dall's and harbor porpoises.

Dall's porpoise are observed inshore and offshore. They prefer deep cooler temperate and subpolar waters of the North Pacific Ocean ranging from Alaska to the Bering Sea and east to Japanese waters. As they swim rapidly at high speed, the heavily muscled tail produces a V-shaped splash called a "rooster tail." They are avid bowriders often seen at the bow of fast moving boats. Other times they move slowly and roll at the surface.

Dall's porpoises are usually found in fluid groups varying in size from two to twelve individuals, but they have been seen in loose aggregations in the hundreds or even thousands especially when feeding. They are known to associate with Pacific white-sided dolphins and short-finned pilot whales as well as swim alongside large whales. Compared to other porpoises, they tend to be found in deeper water at depths of 1,968 feet (600 m) or more. They are opportunistic feeders taking available prey in specific locations including schooling fish, mid and deep water fish, cephalopods, and occasionally crustaceans.

Dall's porpoises have a polygynous mating system in which males compete for females. Research has shown that Dall's porpoise display a rare strategy during the

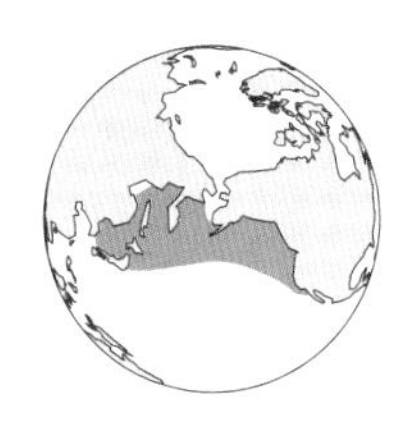

SIZE

Males: length 6 ft–8 ft (1.8 m–2.5 m); weight 271 lb (122 kg)
Females: length 5 ft 6 in–7 ft (1.7–2.2 m); weight 271 lb (122 kg)

DIET

Fish and squid

HABITAT

Nearshore, deep warm temperate-subarctic waters

LIFE HISTORY

Sexual maturity
Males: 3.5–8 years
Females: 4–7 years
Gestation: 10–12 months
Reproduce: yearly

STATUS

Least concern
Threats from hunting, incidental fisheries

A Dall's porpoise showing the distinctive color pattern with a white patch on the belly and flanks. This is the fastest swimming small cetacean, reaching speeds of 34 mph (55 kph) over short distances.

breeding season in which males try to hold onto their partner by mate guarding, whereby a male closely associates with a female. While this behavior is often observed in terrestrial species, it is rare in cetaceans given their wide dispersal and the mobility of females and their prey. This behavior occurs when the male closely attends and defends a female's movement when they are near one another and challenges any potential rivals. While this behavior has obvious benefits since the male increases the changes of mating with the female he guards, there are negatives such as reducing the time the male might otherwise spend feeding. Study of Dall's porpoises that employed mate guarding revealed that males were observed aggressively interacting with other adult males, suggesting the males were defending their mates against rivals.

Large numbers of Dall's porpoise continue to be incidentally caught in Japanese gill net fisheries and coastal whaling operations, and they are also subject to direct kills for human consumption of meat and blubber products.

Dall's porpoise is one of seven species in the family Phocoenidae most closely related to the harbor porpoise.

# West Indian Manatee

## Body hairs scattered over the body in the West Indian manatee may be analogous to the lateral line system of fish used to detect water movement.

West Indian manatees (*Trichechus manatus*) have small heads, rotund bodies, long forelimbs, and a paddle-shaped tail. They are the largest of the three manatee species and females are generally larger than males. The color of the skin is gray to brown sometimes with green, red, white, or black tinges caused by algal or barnacle growth. The rostrum is strongly deflected downward in this species, more so than in the other manatee species, which reflects this species' preference for feeding in submerged vegetation using their fleshy mobile lips. Like other manatees, the West Indian manatee possesses a conveyor-belt dentition: when worn teeth fall out in the front they are replaced by teeth erupting in the rear of the jaw. Likely this is related to their consumption of abrasive grasses that promote increased tooth wear. Manatee bones are dense and solid, which provides ballast and helps them remain horizontal in the water when feeding.

All species of manatees have small eyes and poor vision as well as low-frequency hearing, which likely accounts for their collisions with slow-moving boats that don't provide warning signs of their approach. Their vocalizations consist of squeaks, squeals, and screeches with sex and age-related differences. They are the only mammals known to have sensitive hairs all over their bodies that may be used for navigation, sensing water movements and obstacles. The sensory function of body hairs in manatees may serve a similar function to the lateral line systems of fish that are employed to sense water movement when approaching other animals.

As noted in Chapter 2 (see page 64–67) two subspecies of the West Indian manatee are recognized. The Florida manatee (*T. m. latirostris*) and the Antillean manatee (*T. m. manatus*) are distinguished by geography and differences in skeletal morphology. Florida manatees are found in coastal marine, brackish, and fresh waters along the southeastern US coast. The Antillean manatee is located throughout the Caribbean Sea, including the Gulf of Mexico, and the Atlantic coast of northeastern South America. They feed on a variety of aquatic plants, such as water hyacinths and marine seagrasses. Because of the low-nutrient quality of the vegetation they consume, manatees must graze for 6–8 hours per day. Their diet contributes to their low metabolic rates, among the lowest for a mammal that are sufficient to meet their low energetic demands.

SIZE

Males & females
Length: up to 11 ft–13 ft (3.5 m–4 m)
Weight: up to 1.5 tons (1.6 tonnes)
At birth: length 4 ft (1.2 m); weight 66 lb (30 kg)

DIET

Aquatic plants (water hyacinth, marine seagrasses)

HABITAT

Marine, estuarine and fresh water nearshore and riverine waters

LIFE HISTORY

Sexual maturity
Males and females: 3–4 years
Gestation: 11–13 months
Reproduce: 3–5 years

STATUS

Endangered
Threat from boat collision

RIGHT
Florida manatee (*Trichechus manatus latirostris*) swims in the Crystal River, Florida. The manatee uses its flexible forelimbs to assist in locomotion with propulsion generated by the paddle-like tail.

OVERLEAF
West Indian manatees feeding on seagrasses.

ABOVE
West Indian manatee mother and nursing calf. Manatees suckle a few hours after birth from teats located under the female's foreflipper.

OPPOSITE
Florida manatees gather at Three Sisters Springs, Crystal River, Florida. Manatees often congregate in warmer water during the winter months.

West Indian manatees spend most of their time resting, eating, or traveling. They are mostly seen alone or in small groups of up to six with some feeding groups numbering up to 20 individuals. During cold weather even larger aggregations of hundreds of West Indian manatees can be found near warm water sources, such as power plants. As coal and nuclear power plants are closed in Florida, it is unclear whether warm water refuges in the central and northern part of the state will be sufficient to support current numbers of the animals.

In recent years, Florida's manatees are dying at an alarming rate with more manatees dying in the first half of 2021 than in any other year of Florida's history. This mass dying event was primarily due to starvation, the result of polluted river systems, and a scarcity of seagrasses.

West Indian manatees are in the family Trichechidae and they are most closely related to Amazonian manatees.

# DUGONG

## Dugongs have a more complex social behavior than their closest manatee relatives. Males in some locations engage in competitive fights and displays to attract females.

Dugongs (*Dugong dugon*) are one of four species in the order Sirenia distantly related to elephants, hyraxes, and other relatives. They are large and gray in color. Like manatees they are slow-moving, earning them the common name sea cows.

Dugongs are unique among sirenians for having whale-like flukes instead of the rounded paddle-shaped tail of manatees. Like whales the hind limbs of dugongs are absent. Dugongs swim using an up and down movement of the tail with the assistance of foreflippers. Dugongs spend a considerable amount of time in turbid water and they are often active at night, conditions not conducive to visual cues. They have poor eyesight and communication is through chirp-squeaks during mother-calf interactions and during mating. Like manatees, the ribs and limb bones of dugongs are thick and dense and give them ballast, helping them stay suspended just below the water's surface.

Also differing from manatees, the dugong muzzle is strongly downturned, making them obligate bottom feeders eating primarily seagrasses rich in nutrients and easily digestible, aided by their large, horseshoe-shaped upper lip, called the rostral disc. Feeding trails gouged by the rostral disc and heavy bristles of dugongs along the sea bottom have been observed. The small tusks, possessed by males and some older females, do not extend outside the closed mouth and are involved in mating activities. The relatively low-nutrient and energy content of their aquatic plant food means that they must spend a high proportion of their time feeding in order to meet their daily food requirement.

The diving behavior of dugongs in Shark Bay, Australia, indicates that they are shallow water divers although they are capable of making dives deeper than 65 feet (20 m) and for up to 12 minutes. Approximately two-thirds of all dives are either feeding or bottom resting dives made during daylight hours. Most feeding dives (square and U-shaped) are only a few yards. They have been known to follow rivers for miles from the coast but cannot tolerate water temperatures below 68°F (20°C) for long periods. They are depth dependent on the seagrasses growing in the intertidal and shallow subtidal areas. Less than 10 percent of V-shaped dives were exploratory dives and just under 25 percent of dives were described as traveling dives.

SIZE

Males & females
Length: up to 11 ft (3.4 m)
Weight: 1,257 lb (570 kg)

DIET

Seagrasses

HABITAT

Shallow waters and continental shelf

LIFE HISTORY

Sexual maturity
Males: 10–12 years
Females: 7–17 years
Gestation: 13–14 months
Reproduce: 3–7 years

STATUS

Vulnerable
Threat from habitat destruction

A dugong mother and calf swim in the Indo-Pacific Ocean. A strong bond is formed between mother and calf during the long weaning period.

A dugong showing its distinctive whale-shaped fluke and short, broad flippers.

Dugongs have a large, fragmented range in the Indo-Pacific that extends from East Africa to Vanuatu in the South Pacific. The largest dugong population occupies Australia's northern seas, between Shark Bay and Cape York. Dugongs mostly occur in small groups with the only long lasting social unit being the cow and calf. This is due to the inability of seagrass beds to support large populations, but larger herds of several hundred animals are seen at various locations. Although dugongs are not migratory, they do undertake large-scale movements of more than 250 miles (400 km) in order to find food. Dugongs generally live in tropical and subtropical shallow waters of sheltered bays, mangrove channels, and protected areas on offshore islands.

Dugong populations have been shrinking globally, mostly due to habitat fragmentation and ocean pollution. Other factors contributing to their decline are reduced genetic variability compared to past diversity and low rates of reproduction, giving birth only every 3–7 years. Scientists have used historic DNA to discover some of the highest-risk populations, such as the Indian Ocean dugongs, and especially the Madagascar dugong, a separate genetic lineage. The danger is that if we lose genetically distinct animals they can never be recovered.

Dugongs are known by a single living species, and they are members of the family Dugongidae mostly closely related to manatees, family Trichechidae.

## MATING: MALES DISPLAY AND FEMALES CHOOSE

The mating behavior of dugongs varies depending on location. Some male dugongs in Shark Bay display a mating behavior similar to lekking, seen in some insects, birds, fish, and mammals. In a lek, males put on a show to attract females to enter their territory for courtship. Females visit the territory where males are displaying to select a male and mate with him. Male dugongs elsewhere form mating herds, similar to manatees, and perform stereotypic rushing behavior at the surface of the water followed by attempted mating of a female by rolling her onto her back in the water. Other behaviors related to mating include splashing, following fighting, and mounting. Fights between males results in scars presumably made by the tusks of other males during male-to-male fights in competition for females. Typically, such mating associations are temporary.

# GALAPAGOS FUR SEAL

## To escape high temperatures, Galapagos fur seals take shelter in caves and in the shade of boulders. They also employ "sand flipping" behavior using their foreflippers to flip sand onto their bodies for cooling.

As the name suggests, Galapagos fur seals, *Arctocephalus galapagoensis*, are mostly found in nearshore tropical waters surrounding the Galapagos Islands in the eastern Pacific off the coast of Ecuador, one of the most restricted geographic ranges for a pinniped.

Galapagos fur seals are the smallest, least sexually dimorphic otariid seal. Their small size allows them to lose heat more quickly, an advantage in warm climates. Their cardiovascular system also aids in cooling by sending more blood to the flippers that dispense heat. Adult males are larger with thicker necks and shoulders than females. Galapagos fur seals are born with a black coat that molts to medium to dark brown as they become adults.

They also reduce their activity during the day, increasing their activity at night when there is no danger of overheating. They are nocturnal foragers with large eyes feeding on a variety of fish and small squid that migrate at night near the water's surface, making them easier to capture. They typically remain close to the coastline where they can find shade in rock ledges in the warm climate.

The behavior of Galapagos fur seals has been well studied. Males hold large territories close to feeding areas. Males do not become physically mature and able to compete for a territory that will be used by females until they are older than five years. Colonies are located close to foraging areas.

Galapagos fur seals have a polygynous mating system with males mating with between 6 and 16 females in a single breeding season. There is no evidence of migratory behavior and they do not spend long periods of time at sea. Galapagos seals have one of the lowest reproductive rates reported in seals, requiring an unusually long time to raise seal pups to independence. Weaning varies between 18 to 36 months.

Galapagos fur seal populations have undergone drastic declines related to the nineteenth-century sealing operations; and more recently they have been adversely affected by warming events. Reduced prey availability during El Niño years has caused female fur seals to increase their foraging efforts so the pups receive less energy and their survivability decreases.

Galapagos fur seals are in the family Otariidae and they are most closely related to the Australian fur seal.

SIZE

Males: length 5 ft (1.5 m); weight 141 lb (64 kg)
Females: length 4 ft (1.2 m); weight: 58 lb 8 oz (27 kg)
At birth: weight males 8 lb 3 oz(3.8 kg); females 7 lb 8 oz (3.4 kg)

DIET

Squid, fish

HABITAT

Coastal, tropical waters

LIFE HISTORY

Sexual maturity
Males: 7–10 years
Females: 3–5 years
Gestation: 12 months
Reproduce: yearly

STATUS

Endangered
Threat from habitat degradation; hunted in the past

Galapagos fur seals typically enter the water during the day to cool off.

# 5 ECOLOGY AND CONSERVATION

The ecological roles of sea mammals explored in this chapter include life history parameters and the effects of environmental change on ecosystem dynamics. Also considered are threats, conservation status, and the actions taken to protect species.

The North Atlantic right whale—although recovered from commercial whaling—is the most endangered baleen whale today. The gray whale is a conservation success: this species has been removed from the Endangered Species list, and may recolonize the Atlantic in the future due to climate warming.

Among toothed whales, time has run out for the Yangtze river dolphin; it is almost certainly extinct and has had no recent sightings. The vaquita (gulf porpoise) is likely the next species headed for extinction due to human activities. In the past, the pantropical spotted dolphin's affinity for associating with tuna led to its entrapment in fishing nets; however, following a change in net design, there has been a marked decrease in mortality. The once highly endangered Hector's dolphin is making a comeback, the result of the establishment of a sanctuary where gillnet fishing is banned.

Historically, two pinnipeds—northern elephant seals and northern fur seals—were hunted to near extinction in the North Pacific; however, populations of both species have

PAGE 166
A sea otter (*Enhydra lutris*) highlighting the dense fur covering its body.

rebounded. Northern elephant seals went through a severe population bottleneck, which resulted in a marked reduction of genetic variability that makes them vulnerable to disease. Habitat disturbance is the major threat to the world's rarest seal, the Mediterranean monk seal. Although harp seal hunting has declined, this Arctic pinniped continues to be killed in Canada. Reduction of the western population of northern sea lions may be the result of changes in prey distribution and availability, and killer whale predation.

Sea cows today occupy tropical and subtropical waters, but Steller's sea cow inhabited cold Arctic waters before being hunted to extinction. Historical and current hunting are responsible for the endangered status of the Amazonian manatee.

The sea otter and sea mink in the carnivoran family Mustelidae were hunted for their fur and meat. The world's largest bear, the polar bear, an iconic symbol of global warming, is in danger of becoming extinct by the end of the twenty-first century.

# VAQUITA

## The vaquita is the world's most endangered sea mammal; there are fewer than ten individuals remaining, and the population has declined by nearly 99 percent since 2011.

SIZE

Males: length 5ft (1.5 m)
Females: length 5ft (1.5 m)
Both sexes: weight 60 lb–150 lb (27 kg–68 kg)
At birth: length ca. 2 ft (0.6 m); weight: 16 lb–22 lb (7.5 kg–10 kg)

DIET

Fish and squid

HABITAT

Coastal

LIFE HISTORY

Sexual maturity
Females & males: 3–6 years
Gestation: 10–11 months
Reproduce: every 2 years

STATUS

Critically endangered
Threat from incidental fisheries catch

Vaquitas (*Phocoena sinus*) were first discovered in 1958, yet less than half a century later we are on the brink of losing them forever. The name *vaquita* means "little cow" in Spanish. Vaquitas are one of the smallest cetaceans and females are larger than males. Vaquitas have a distinctive color pattern. There are conspicuous dark rings around the eyes and dark patches on the lips that form a thin line from the lips to the flippers. Compared to other porpoises, vaquitas exhibit relatively large dorsal fins, flippers, and flukes. The larger dorsal fin and flippers help the vaquita release excess heat. Vaquitas are polydactylous, meaning that they have an extra digit in each flipper, which may be the result of genetic drift in a small population. As is true for other porpoises, vaquitas have spade-shaped teeth and they feed on a variety of fish, squid, and crustaceans.

Vaquitas have a very restricted distribution occurring only in the upper Gulf of California between Baja California and mainland Mexico. Based on its close relationship to Southern Hemisphere porpoise species, the similar timing of rapid climate warming, and vaquita decline, it has been hypothesized that climate change at the end of an ice age caused a northward shift of the species range, resulting in a remnant population being isolated in the Gulf of California. They prefer turbid waters 66–131 feet (20–40 m) deep. While most porpoises inhabit cold waters, water temperature in the vaquita habitat can exceed 90°F (32°C) in the summer and fall. Vaquitas are shy and elusive and occur in small groups, generally no more than two individuals, often mother-calf pairs, but loose aggregations of up to eight to ten are known. They do not display aerial behavior, preferring to spend most of their time under water.

Today, the vaquita has the unfortunate distinction of being the most endangered of the world's sea mammals. Vaquitas are frequently trapped and as a result drown in gillnets, used by illegal fishing operations set up to catch the endangered totoaba fish. Gillnets, also known as "curtains of death," hang upright below the water's surface up to 20 feet (6 m) deep and stretch the length of several football fields; these enormous nets also entangle fish and shrimp. The high demand for totoaba swim bladders has been driven by a perceived Chinese belief that they have medicinal value. Although international trade in totoaba is banned, swim bladders are dried, smuggled, and sold outside of Mexico. They can

Vaquitas can be distinguished from other porpoises by their relatively taller dorsal fin.

command very high prices on the black market; according to a 2018 report, just a gram of it can sell for $46. Other potential threats to the vaquita include pollution, and ecological changes as the result of reduced flow from the Colorado River.

Vaquitas are believed to have a polygynous mating system, where sperm competition plays an important role inferred from the large size of the testes. Sperm competition is a reproductive strategy that occurs when two or more males copulate with the same female, and their sperm competes within the female's reproductive tract for fertilization.

Genetic analyses and population simulation analyses indicate that the vaquita has always been rare and that its low genetic diversity occurred over time rather than being caused by human activities. One study has found little evidence of inbreeding or other risks associated with small populations. The implication is that despite the current low population the removal of human threats to the species, that is, the elimination of all fishing deaths would change the outcome of the species from it being headed toward extinction to its survival.

The vaquita is a member of the family Phocoenidae and is most closely related to Burmeister's and spectacled porpoises, which are only found in the temperate and cold waters of the Southern Hemisphere.

## THE BATTLE OVER FISHING NETS

Various vaquita conservation efforts have been carried out over the last decades. In 2015 the Mexican government enacted a two-year ban on gillnet fishing in the vaquitas' habitat; this was followed a year later by a permanent ban. However, enforcement of the ban proved difficult, and it is illegal fishing that is largely responsible for the vaquita's dramatic decline. Other efforts to save the vaquita enacted by the Mexican government included expansion of the protected area for vaquitas to encompass their entire range, and compensation to the fishing communities affected by the ban. Sadly, none of these measures worked and in 2021 the Mexican government lifted the ban on gillnet fishing—a serious setback for the vaquitas' survival. They also eradicated a "no tolerance" zone in the Upper Gulf of California meant to protect the vaquitas' habitat and opened it up to fishing.

In 2017 conservationists and the Mexican government attempted an ambitious plan to save the vaquita by bringing some into captivity. The plan involved using trained US Navy dolphins to locate vaquitas, which would then be temporarily relocated to large floating sea pens in an ocean sanctuary to protect them. The plan was abandoned because the porpoises became so stressed by contact with humans that a captured female died and a second captured female showed signs of stress and had to be released.

In another effort to save the vaquita in 2020 the US banned seafood caught from the vaquita's habitat in the upper Gulf of California using gillnets. With fishermen losing access to US markets, Mexican authorities are under pressure to step up the enforcement of gillnet bans and the development of alternative vaquita-safe sustainable fishing gear.

Conservation actions implemented to protect the vaquita have not led to an increase in numbers of vaquitas. Despite policy changes and considerable financial investments that were made by the Mexican government, they have been unsuccessful in eliminating the threat that gillnets pose to this species. A key missing component was an inability to prioritize and support alternative fishing methods and livelihoods for coastal communities. These issues were made more difficult by the failure to address fisheries management and governance problems. Underlying these concerns was lack of community buy-in. Ultimately, conservation measures are likely to be more successful if they are inclusive and benefit local communities using multidisciplinary expertise and a broad range of tools and policy instruments.

ABOVE
As a vaquita surfaces the dark eye rings are revealed.

RIGHT
Vaquitas are often entangled in fishing nets, which can result in death.

# Amazonian Manatee

## The Amazonian manatee is the smallest and most endangered of three species of manatee and the only exclusively freshwater species.

The Amazon river manatee (*Trichechus inunguis*), one of five species in the order Sirenia, is found in Brazil, Columbia, Peru, Guyana, and Ecuador. It inhabits the warm, shallow waters of the Amazon River and its tributaries, preferring lakes, oxbows, and lagoons that connect to large rivers. It is the largest mammal occupying South American fresh waters. Hybrids between West Indian and Amazonian manatees have been documented near the mouth of the Amazon River and probably also near the mouth of the Orinoco River. Distributional information is scarce since they are visually very difficult to detect; they only briefly expose their snout when they breathe at the water's surface.

Amazonian manatees are black or dark gray, and most have a white or pink belly and chest patches. They can be distinguished from other manatee species by the lack of nails on their flippers and smoother rubbery skin.

Although once known to occur in large herds, today the Amazonian manatee occurs singly or in feeding groups of up to eight individuals. They generally feed at the surface on a variety of floating aquatic plants, including grasses, water hyacinths, and floating palm fruits. They make seasonal movements that are synchronized with the flood regime of the Amazon River and availability of food. They feed mostly during the wet season, eating new vegetation, and they rely on stored fat reserves during the dry season until food becomes available again. They consume approximately 8 percent of their body weight in food per day.

Amazonian manatees are docile, slow moving, found at the water's surface, and easy to hunt, which has contributed to their vulnerable status. They are victims of historical and current hunting, pollution, and habitat degradation. They have long been hunted by native peoples as well as European colonizers, who sought them for their leather, meat, and fat. They continue to be hunted today for subsistence by local communities and for illegal trading purposes.

In 2003, one of the largest conservation projects began in Anavilhanas National Park in the lower Rio Negro region of Brazil. Areas used by manatees were monitored, principal threats were identified, and conservation programs were developed for the species and its habitat. Results revealed that public awareness was very important—engaging local populations about species conservation and connecting them to project activities. Another finding

SIZE

Males & females
Length: 9 ft–10 ft (2.7 m–3 m)
Weight: up to 1,058 lb (480 kg)
At birth: length 4 ft–4ft 6 in (1.2m–1.37 m); weight 1 lb 12 oz–2 lb 7 oz (0.8 kg–1.1 kg)

DIET

Aquatic plants

HABITAT

Coastal, estuarine

LIFE HISTORY

Sexual maturity
Males and females: 6–10 years
Gestation: 12 months
Reproduce: 2–5 years

STATUS

Vulnerable

The distinctive white belly of the Amazonian manatee is a distinguishing feature of this species.

was that the more the local communities knew about the Amazonian manatee—its biology, ecology, and threats—the more likely they were to become actively involved in its conservation.

The Amazonian manatee is most closely related to the West Indian manatee and both in addition to a third species, the African manatee, are in the family Trichechidae.

# Mediterranean Monk Seal

## The Mediterranean monk seal is the world's rarest seal—only 600–700 individuals remain. This species is a victim of human hunting and habitat disturbance.

Mediterranean monk seals (*Monachus monachus*) are one of two surviving species of monk seals, and one of a few pinniped species to occupy subtropical and tropical waters. They inhabit sandy beaches and shoreline rocks and can seek refuge in inaccessible caves. Once found throughout the Mediterranean Sea, and in parts of the Atlantic and Black Sea, Mediterranean monk seals are scattered in three isolated populations across a portion of their former range: Eastern Mediterranean, Madeira archipelago, and Mauritania. A second genus of monk seals *Neomonachus* is recognized as a New World species based on molecular and morphologic data. The Hawaiian monk seal (*Neomonachus schauinslandi*) occupies the Hawaiian Islands, and like the Mediterranean monk seal is endangered although there are signs of a population increase over the last few years. A third species, the Caribbean monk seal (*Neomonachus tropicalis*), was widespread in the Caribbean Sea, Gulf of Mexico, and West Atlantic Ocean but was hunted to extinction by humans in the twentieth century.

Mediterranean monk seals are sexually dimorphic. At birth pups are born with black woolly coats with a white patch on the belly. Males have a white belly patch and are black or dark brown, and resemble the robe of a monk, hence the name. Females have a lighter belly and are generally brown or gray. Other irregular patches are sometimes found mainly on the throats of males and backs of females, often the result of scarring sustained in social and mating interactions.

Mediterranean monk seals do not migrate; most individuals spend much of their time within a limited home range. Although once a gregarious species forming colonies, they are now more often solitary and reclusive, but can occasionally occur in larger groups from 20–200 individuals. Pup survival during the first two months of life is low due to storms, large ocean swells, and high tides.

Mediterranean monk seals exhibit a unique breeding behavior in which females seek out new pupping sites due to limited space in existing ones, and males need to defend aquatic territories around pupping sites. The female remains with the pup after giving birth, living off stored fat. As a result, in some subpopulations there is little dispersal and the species survives in fragmented small populations

SIZE

Males: length 8 ft 6 in (2.6 m); weight 529 lb–661 lb (240 kg–300 kg)
Females: length 8 ft (2.4 m); weight 529 lb–661 lb (240 kg–300 kg)

DIET

Bony fish, cephalopods (mainly octopus), crustaceans

HABITAT

Tropical and subtropical waters

LIFE HISTORY

Sexual maturity
Males: 7 years
Females: 5–6 years
Gestation: 9–11 months
Reproduce: yearly

STATUS

Endangered
Threats from human disturbance, habitat destruction

A juvenile male Mediterranean Monk seal located at the Desertas Islands, Madeira, Portugal; this species is critically endangered.

that have low genetic diversity, and high levels of inbreeding, which increases their risk of extinction. Extant populations show almost a four-fold decrease in genetic diversity compared to historic specimens. Such a loss in genetic diversity highlights the need to implement effective conservation strategies, such as a reduction in habitat loss to save this species from extinction.

Mediterranean monk seals dive and feed in shallow, nearshore waters. They are "opportunistic" feeders consuming a variety of fish including eels, sardines, tuna, crustaceans, and cephalopods, such as octopuses.

Mediterranean monk seals have a long history of human exploitation. Conservation efforts emphasize tagging and monitoring of seals as well as the creation of protected areas. In one locale in Turkey artificial ledges constructed in caves allowed the successful breeding of monk seals.

Mediterranean monk seals are most closely related to New World monk seals (Caribbean and Hawaiian monk seals) and they are members of the family Phocidae, subfamily Monachinae.

# SEA OTTER

## The luxuriant pelt of the sea otter, the densest fur of any sea mammal, led to its near extinction after centuries of exploitation by the fur trade.

Three subspecies of sea otter (*Enhydra lutris)* are recognized based on morphologic and geographic distributions. The common sea otter (*E. l. kenyoni*) inhabits islands off Japan along the Pacific coast, and the northern sea otter (*E. l. lutris*) ranges from Alaska to Oregon. The southern sea otter (*E. l. nereis*) had a historic range from northern California to central Baja California, Mexico, and is present today along the California coast. Historically, sea otters numbered more than 300,000 individuals. The maritime fur trade during the eighteenth and nineteenth centuries reduced sea otter populations to less than 2,000 individuals. The first legal protection for sea otters was in 1911 with enactment of the North Pacific Fur Seal Treaty. However, their pre-exploitation range had already been seriously fragmented with widely scattered remnant populations. In both Canada and the US, sea otters are protected under federal laws. Although full reoccupation of all historical habitat has not occurred, many populations have begun to recover.

The ability of sea otters to live in cold waters has long been known, but less understood until recently was how these small marine mammals—males average 4 feet (1.2 m) in length—managed to maintain a metabolism equal to mammals three times their size. Scientists demonstrated that the skeletal muscles of sea otters are well suited to generate heat, which helps them to survive living in cold water. Specifically, the mitochondria, tiny organelles that produce energy in the cells, are leaked into their skeletal muscle increasing their respiratory metabolism. The primary form of insulation for sea otters is an exceptionally thick coat of fur, with up to 1 million hairs per square inch.

Sea otters can be seen singly or in resting groups of 2–12 individuals, known as rafts. Sea otters are most often seen floating on their backs near kelp beds. When feeding they bring their prey to the surface and use rocks as tools to crack open the shells of their prey. They have a diverse diet and prey on abalones, sea urchins, and other shellfish.

The mating system of sea otters is resource defense polygyny. Females and pups occupy areas that provide protection and distribution of food resources (i.e., kelp). Adult males establish territories within these areas and actively defend them, limiting the number of males in the areas with females moving freely among territories.

SIZE

Males: length 5 ft (1.5 m); weight 99 lb (45 kg)
Females: length 2 ft (0.6 m); weight 4 lb–4 lb 4 oz (1.8 kg–1.9 kg)
At birth: length 2 ft (0.6m); weight 3 lb–5 lb (1 kg–2 kg)

DIET

Benthic invertebrates (mostly sea urchins, abalones, rock crabs)

HABITAT

Nearshore rocky coasts

LIFE HISTORY

Sexual maturity
Males: 4–6 years
Females: 3–5 years
Gestation: 6–7 months
Reproduce: yearly

STATUS

Endangered
Threat from climate change; hunted in the past

RIGHT
Sea otters are not common on land, although they can walk on all four feet.

BELOW
Sea otters often congregate in resting groups or rafts that can contain as many as 2,000 otters.

Sea otters are considered keystone predators and they play a pivotal role by keeping sea urchin populations, which in turn feed on kelp, low in number. Some sea otter populations, notably those in the Aleutian Islands, declined in the 1990s as a result of killer whale predation. Again, this reflects a change in ecosystem dynamics, in this case driven by the decline in whaling and prey switching by orcas from whales to seals and sea otters.

Sea otters are the largest members of the family Mustelidae, subfamily Lutrinae.

## TRANSLOCATING SEA OTTERS

Sea otter reintroduction projects were undertaken during the twentieth and twenty-first centuries; they involved moving sea otters from a few healthy, remaining populations to parts of their historic range. One of the best-known translocations was in the late 1980s when the US Fish and Wildlife Service relocated 139 southern sea otters from the mainland population near Monterey Bay, California to San Nicholas Island, one of the Channel Islands, approximately 65 miles (105 km) offshore. The goal was to exclude otters from their original habitat—a so-called "No Otter Zone"—by relocating a colony of sea otters, so that they would not be affected by a nearshore catastrophic oil spill, and would not deplete the commercial shellfish fishery off the coast of southern California. The translocated animals had very low survival rates. Many otters swam back to their waters of origin; others died from being captured or transported. But, lessons were learned. The homing behavior of the sea otters was not adequately considered nor was the high mortality rate following the animals' release.

The success of translocations was studied by investigating changes in genetic diversity over time among translocated populations. High rates of genetic diversity were found in some translocated populations in northern Washington, central British Columbia, and southeastern Alaska. This has been attributed to these areas possessing plentiful space for populations to grow and abundant resources to exploit. Other populations, including those in California, showed low genetic diversity. Recolonizing coastal and estuarine habitats in these areas would fill the distributional gap between Washington and California, enhancing gene flow and restoring genetic diversity. Despite challenges the US Fish and Wildlife Service is currently reevaluating the feasibility of sea otter reintroduction along parts of the Pacific coast.

OPPOSITE TOP
Sea otters use their forepaws to manipulate bottom-dwelling invertebrate prey, which they then crush using their massive jaws and teeth.

OPPOSITE BOTTOM
As female sea otter protectively holds her newborn pup out of the water, located in Prince William Sound, Alaska.

# STELLER'S SEA COW

## Steller's sea cow has the dismal distinction of becoming extinct less than three decades after its discovery.

SIZE

Males & females
Length: at least
24 ft 7 in (7.5 m)
Weight:
4.40 tons–11 tons
(4 tonnes–10 tonnes)

DIET

Algae, especially kelp

HABITAT

Cold shallow coastal waters

STATUS

Extinct in historic times (1768)

Steller's sea cow (*Hydrodamalis gigas*) was adapted to living in cold waters unlike their close relative the dugong that today lives exclusively in tropical waters. They were large, heavy animals living in coastal waters around the North Pacific, from Mexico to Alaska to Japan. They were quickly hunted to extinction by sailors and fur traders for their meat and fat. Named for their discoverer, Georg Steller, a German naturalist, Steller's sea cow blubber was described as tasting like almond oil, which likely contributed to their demise. The disappearance of Steller's sea cow has been regarded as the first historical extinction of a marine mammal as a consequence of human action.

Steller's sea cows possessed a thick, rough skin that was described by Steller as "a skin so thick that it is more like the bark of old oaks than the skin of an animal." Study of the genome of Steller's sea cow reveal changes that were responsible for the bark-like skin and the adaptation to cold. Several genes reveal an increased evolutionary rate in this species that resulted in an increased capacity for fasting and fat accumulation, which may have been key to their survival in cold waters. The genome also reveals important insights about present day skin diseases in humans.

Steller's sea cows were toothless. Instead of teeth they had large horny pads that were well suited for eating kelp. The predominant kelp species in the western North Pacific is only seasonally present so that the animals would have needed to fast for many months each year. Their bones were very dense to help counteract the problems faced with buoyancy, due to their thick blubber. They pulled themselves through the water with their short forelimbs using their flukes for propulsion. Another unique feature was their lack of finger bones, retaining only a hook-like forelimb, reduced forelimb muscles and modified elbow joints that permitted them to "walk" in shallow substrates when feeding. Due to their great buoyancy, they had difficulty submerging and floated with their backs exposed. They congregated in small herds in shallow waters with young at the front surrounded by adults.

Steller's sea cow influenced the dynamics of kelp forests across the northern Pacific Ocean. The consumption of the kelp canopy would have allowed more sunlight to reach the understory benefiting kelps growing below. Grazing by sea

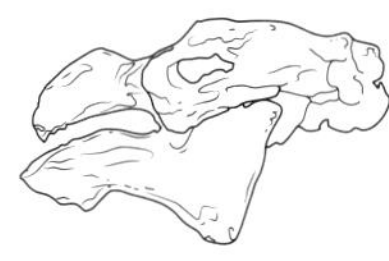

Steller's sea cow had bark-like skin and stubby forelimbs. Its skull is distinguished by its large bony relatively undeflected snout, and lack of teeth.

cows could also have affected the dispersal of kelp, spores, and nutrients within the ecosystem and to other parts of the ocean.

The extinction of Steller's sea cow was probably a consequence of the co-occurring losses of kelp and sea otters. Knowledge of the importance of sea otter-kelp interactions and a study of the living dugong's response to food loss, revealed that sea cows living around Bering Island (the supposed location of the last population of Steller's sea cow) would have reached near or complete extinction by the end of 1768, the year of their last reported sighting. Although human hunting may have contributed to the decline of Steller's sea cow at Bering Island, its extinction was also a nearly inevitable consequence of the loss of sea otters and kelp forests and would have occurred without human interference. Study of the Steller's sea cow genome confirms that extinction of the species began well before the arrival of humans along the North Pacific coastline.

Steller's sea cows are members of the family Dugongidae.

# NORTH ATLANTIC RIGHT WHALE

## North Atlantic right whales, having rebounded from near extinction during commercial whaling, have seen a 30 percent population decline in the last decade.

SIZE

Males: length up to 55 ft 9 in (17 m)
Females: length slightly larger
Weight: 99 tons (90 tonnes)

DIET

Calanoid copepods, smaller invertebrates (copepods, krill, pteropods, larval barnacles)

HABITAT

Atlantic coastal waters

LIFE HISTORY

Sexual maturity
Males & females: 5–10 years
Gestation: 12 months
Reproduce: 3–4 years

STATUS

Critically endangered
Threat from vessel collisions; in the past commercially hunted

North Atlantic right whales (*Eubalaena glacialis*) received their name from whalers, which considered them the "right" whale to hunt because they swim close to shore, float when dead, and produce a large amount of oil when harvested. Once common along both sides of the Atlantic, the North Atlantic right whale was hunted to near extinction by the mid-1700s. A single remaining population occurs along a portion of its former range extending from the Gulf of St. Lawrence, Canada, to the coast of Florida. They are disappearing from the North Atlantic with fewer than 400 individuals remaining. While the population is recovering, these whales continue to be threatened and are victims of ship strikes and entanglements in fishing gear. New technologies that aim to decrease right whale entanglements include ropeless fishing gear—the use of gear without fixed buoy ropes in the water column—that could dramatically reduce or eliminate right whale entanglements while allowing fishing to continue.

This species is distinguished by having stocky bodies, no dorsal fins, and short paddle-shaped flippers. Up to one-third of their body length is taken up by their massive head. They have a long, curved mouthline that supports a massive rack of 200–270 long thin baleen plates. Each baleen plate is up to 8 feet (3 m) long. Their heads are encrusted with white patches of skin called callosities. These callosities are unique and can be used to identify individual whales. Dense colonies of cyamids or whale lice live on the callosities. North Atlantic right whales have also been found to possess sensory hairs on the rostrum and chin that may serve as foraging cues.

Right whales travel in the North Atlantic most often singly or in pairs, or in small groups of less than a dozen. They are strongly migratory, moving annually from high latitude feeding grounds in the Gulf of Maine and low latitude calving grounds in coastal waters off the southeastern US. Larger aggregations may form for feeding and breeding. Their mating system involves sperm competition in which males compete to mate with females by producing large amounts of sperm to displace the sperm of other males.

Right whales feed by skim feeding, slowly swimming through the water with open mouths filter-feeding dense patches of zooplankton, mostly copepods. Right whales, like most other baleen whales, take advantage of the increased

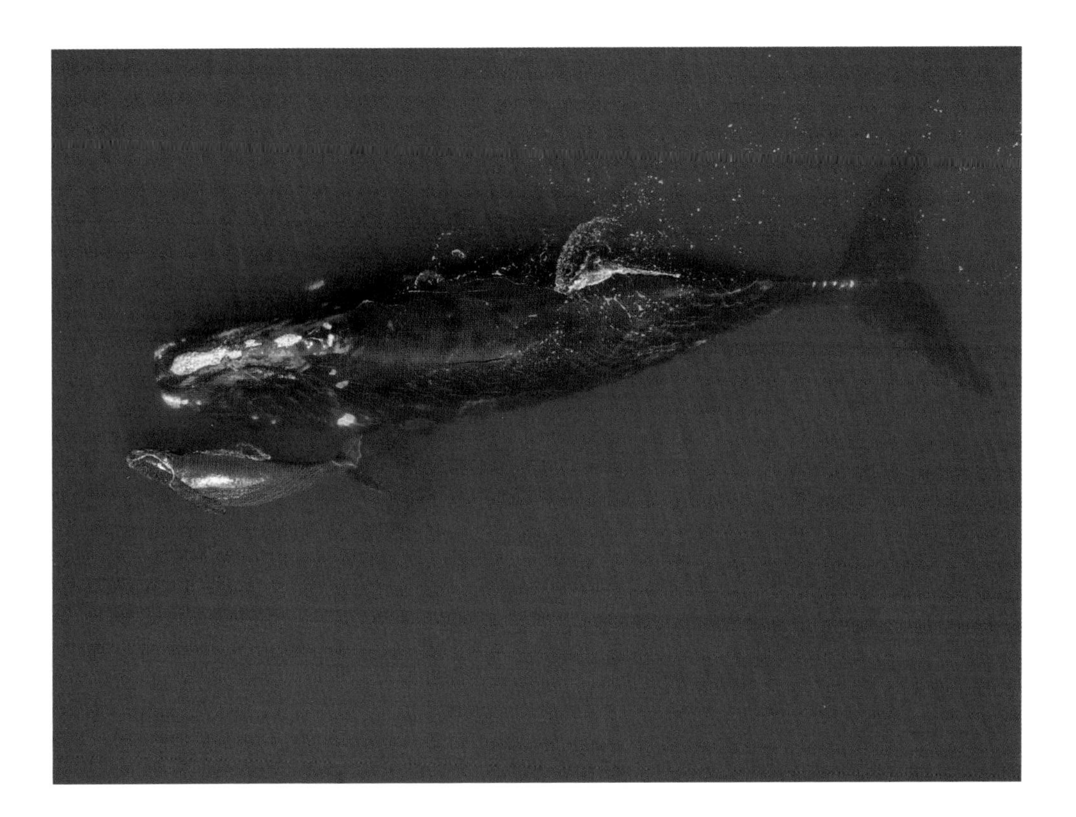

North Atlantic right whale mother swims with her calf alongside. Callosities on the head of the adult whale are a useful identification feature.

biomass available at lower levels in the food chain. However, they are subject to fluctuations in prey caused by environmental variability. When zooplankton populations decline, right whales do not appear capable of switching to an alternative food resource, which further affects their population dynamics and may ultimately govern their survival.

Right whales, like other baleen whales, communicate using low-frequency sounds consisting of moans, groans, belches, and pulses. One type of vocalization termed the "up call" is a short "whoop" sound that may function as a signal that brings whales together.

A study using drone technology monitored the health of North Atlantic whales. It revealed that these whales are struggling to survive, and those living today are significantly shorter in length than those born 30–40 years ago. Scientists suggest that the stunted growth of the whales may lead to reduced reproductive success. Another study confirmed the low reproductive rates of North Atlantic right whales as well as highly variable calving intervals. The species also has a low level of genetic diversity as a result of years of commercial whaling.

The North Atlantic right whale is one of three species of right whale; its closest relatives are the southern right whale and North Pacific right whale, all members of the family Balaenidae.

# NORTHERN FUR SEAL

## Northern fur seals are another seal that was commercially harvested for their luxurious pelts resulting in severely reduced populations.

The scientific name for northern fur seals, *Callorhinus ursinus*, means "bear-like," which reflects the name originally given by Europeans. Northern fur seals show extreme sexual dimorphism and males are considerably larger than females. Adult males are stocky with an enlarged thick neck. They are black, dark brown, or cinnamon, and females are light brown with lighter colored hairs on the throat and chest. Uniquely in this species, relative to other otariids, fur is absent on the top of the foreflippers, and the hind flippers are proportionately the longest of any otariid. There have been confirmed hybrids of northern fur seals and the California sea lion.

Northern fur seals have a wide geographic range and are found throughout the North Pacific Ocean, ranging from the Bering Sea to southern California; about 45 percent of them live on Alaska's Pribilof Islands. The southernmost breeding colony in San Miguel Island was successfully recolonized in the 1960s, after the local population was extirpated by hunting. They are highly migratory species; some males migrate as far south as the Gulf of Alaska in August while others remain in the Bering Sea. The largest population is found on the Pribilof Islands in the Bering Sea; it has undergone a marked decline since the 1990s, as a result of several factors. For example, commercial fisheries have removed pollock, a principal food source for northern fur seals, and this has been linked with their population decline, as nursing females rely on pollock to feed to their pups.

Northern fur seals are one of the most pelagic pinnipeds: they spend 80 percent of their time foraging at sea, feeding mostly at night on schooling fish (especially walleye pollock) and squid. They forage relatively far from shore near the continental slope and shelf. Foraging dives occur at dusk and dawn; the dives average less than 328 feet (100 m) and last for several minutes. They spend the rest of their time resting, sleeping, or grooming at the water's surface.

Northern fur seals are highly polygynous. Males arrive at the rookeries before females and vocalize, fight, and display to establish and maintain territories. Males do not become large enough to compete for a territory until they are 8 to 9 years old. Females that breed at the Pribilof Islands maker longer foraging trips than most other otariid seals: a trip can last 6 to 9 days. At sea they are usually observed singly or in pairs: groups of three or more are uncommon.

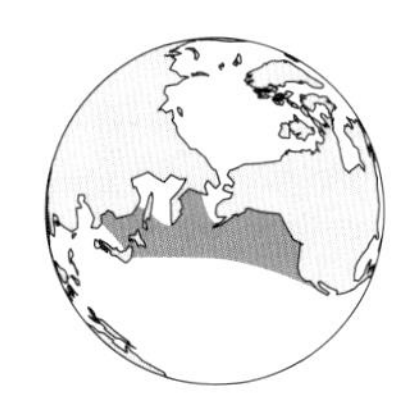

SIZE

Males: length up to 7 ft (2.1 m); weight 595 lb (270 kg)
Females: length up to 5 ft (1.5 m); weight 110 lb 4 oz (50 kg)
At birth: length 2 ft (0.6 m); weight 12 lb–13 lb 4 oz (5.4 kg–6 kg)

DIET

Fish (capelin, herring, rockfish)

HABITAT

Off continental shelf in deep water

LIFE HISTORY

Sexual maturity
Males and females: 3–5 years
Gestation: 12 months
Reproduce: yearly

STATUS

Vulnerable
Threat from fishery interactions; in the past commercially hunted

An adult male (left) with female northern fur seals highlights the distinct difference in size between the sexes. The thick fur and short, blunt muzzle are characteristic features of this species.

This fur seal species was commercially hunted for its fur until enactment of the 1911 International Fur Seal Treaty, which is regarded as the first international treaty for wildlife preservation. Despite this protection and the Marine Mammal Protection Act in 1972, northern fur seal populations have continued to decline. The causes for the more recent declines including reduced prey quality and availability, which is affected by commercial fishing and environmental change, such as increasing temperature.

Northern fur seals are the earliest diverging extant members of the family Otariidae. The genus *Callorhinus* likely evolved in the North Pacific and it is known from California (Plio-Pleistocene) and Japan (Pliocene).

# NORTHERN ELEPHANT SEAL

## During the late 1800s, northern elephant seals were hunted to near extinction. Entire herds in California were slaughtered for the high oil content of their blubber and meat.

Northern elephant seals (*Mirounga angustirostris*) are the largest seal in the Northern Hemisphere. They are countershaded dark gray or brown above and silver gray below, the coloration fades as they age. Pups are black until they are weaned at about six weeks old and develop an adult coloration. Northern elephant seals are found on the coast and offshore islands of the eastern and central North Pacific. Breeding occurs on islands and mainland sites as far north as the Farallon islands and as far south as Baja California. Northern elephant seals exhibit pronounced sexual dimorphism in size and secondary sexual characteristics, such as a thickened chest shield, large canines, and an enlarged nose. In prime breeding condition, males are five to six times larger than adult females. Much of their bulk is composed of blubber, an insulating layer of fat that keeps them warm in cold waters. It is also this blubber oil that nearly sent this species to extinction. Together with their cousins, southern elephant seals that live in sub-Antarctic and Antarctic waters, they are the largest marine mammals other than whales.

Elephant seals are highly polygynous with males competing in dominance battles for access to females. When males inflate their prominent nose, which resembles an elephant's trunk, it hangs down in front of the mouth. During dominance battles, loud snorts and other vocalizations are deflected downward into an open mouth that acts as a resonating chamber, amplifying the sound. The most dominant (or alpha) male in a hierarchy defends nearby females against incursions by subordinate males.

At Año Nuevo on the central California coast, northern elephant seal breeding behavior has been studied for more than five decades. In areas crowded with females, an alpha male can control access to as many as 50 females. Fewer than 10 percent of males manage to mate during their lifetimes, whereas very successful males mate with 100 or more females. A recent study based on four decades of data from the colony at Año Nuevo revealed that the single most important factor in determining the weaning mass of pups, and hence their survival, was the age of the mother. Older mothers produce milk with a higher fat content at the start of lactation, and a more experienced mother is able to select a better location on the beach for rearing her pup. Northern elephant seal milk is a whopping 55 percent fat while cow's milk is 4 percent fat at most.

SIZE

Males: length 14 ft (4.3 m); weight up to 2¾ tons (2.5 tonnes)
Females: length 8 ft (2.5 m); weight 662 lb–1,323 lb (300 kg–600 kg)

DIET

Squid and fish (Pacific whiting, rockfish, ratfish, small sharks, and rays)

HABITAT

Migratory

LIFE HISTORY

Sexual maturity
Males: 6 years
Females: 4 years
Gestation: 11 months
Reproduce: yearly

STATUS

Least concern

ABOVE
A harem of female northern elephant seals watch a larger male, which is readily identified by his large nose, in the San Benito Islands off the west coast of Baja California, Mexico.

RIGHT
A female northern elephant seal nurses a pup at Point Piedras Blancas, California. Normally such seals will nurse for about a month before heading out to sea to feed.

# PANTROPICAL SPOTTED DOLPHIN

## Historically, millions of pantropical spotted dolphins were entangled and killed in nets during the tuna purse seine fishery operations in the 1960s.

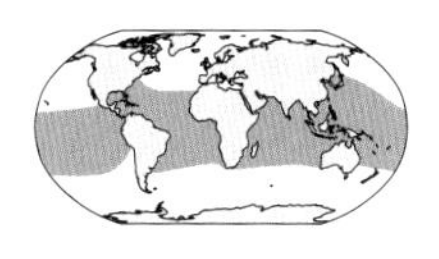

SIZE

Offshore males & females
Length: 5 ft–8 ft (1.6m–2.4 m)
Weight: 262 lb (119 kg)
Inshore adult males & females
Length: 6 ft–8 ft 6 in (1.8 m–2.6 m)
Weight: up to 263 lb (119 kg)

DIET

Small fish, squid, and crustaceans

HABITAT

Mostly offshore in subtropical–tropical waters, some forms are coastal and inshore

LIFE HISTORY

Sexual maturity
Males: 12–15 years
Females: 9–11 years
Gestation: 11 months
Reproduce: 2–3 years

STATUS

Least concern
Threat in the past as fisheries bycatch

Pantropical spotted dolphins (*Stenella attenuata*) are recognized by a dark dorsal cape positioned above the flipper that sweeps along the back terminating behind the dorsal fin. Pantropical spotted dolphins are born spotless. They are named for their light spots that develop as they grow, appearing on their dark-colored backs and have dark spots on their belly and sides. The cape is spotted to varying degrees, less so in offshore animals. Two subspecies are recognized: *S. a. attenuata*, the offshore spotted dolphin, which has a global distribution, and *S. a. graffmani*, the coastal spotted dolphin, which is found only along the Pacific coast of Mexico and Central America.

Pantropical spotted dolphins are one of the most numerous dolphins in the Eastern Tropical Pacific. The two subspecies vary in habitats—inshore and offshore forms live in subtropical and tropical waters in the Atlantic, Pacific, and Indian Oceans. These closely related dolphins are sometimes difficult to distinguish from other species with which they share their range.

Pantropical spotted dolphins feed on a variety of small schooling fish and squid, which brings them into contact with various species of tuna that feed on the same prey. They reside in shallower water during the day and move into deeper water at night to feed.

The spotted dolphin is highly social; they swim in schools of varying sizes from several individuals to several thousand dolphins. They are gregarious and often acrobatic, performing breaches, side slaps, and frequently bow riding. In the Eastern Tropical Pacific they often swim with other dolphin species, such as rough-toothed dolphins, short-finned pilot whales, and spinner dolphins. Like other delphinids, they produce echolocation sounds as well as signature whistles; the latter are at a higher frequency than other *Stenella* species.

For reasons unknown, yellowfin tuna and several dolphin species (see page 146) travel together in the Eastern Tropical Pacific and nets can encircle both tuna and dolphins. Purse seine nets are long walls of netting that are set and used to encircle and trap fish, since the bottom of the netting can be pulled closed like a drawstring purse. The number of dolphins killed since the fishery began in the late 1960s is estimated to be over 6 million animals, the highest known for any fishery. Improvement in net design has substantially decreased the number of spotted dolphins killed.

A pantropical spotted dolphin leaps and reveals its distinctive dorsal cape.

Spotted dolphins are in the family Delphinidae, subfamily Delphininae. Recent molecular work indicates that *Stenella* is not a natural group as it includes the common ancestor and some, but not all, descendants. There are four other genera—*Sousa, Tursiops, Delphinus*, and *Lagenodelphis*—nested with *Stenella*.

# Yangtze River Dolphin

## The Yangtze river dolphin or baiji, *Lipotes vexillifer*, is now believed to be the first dolphin species driven to extinction due to human impacts.

Baijis, one of five species of river dolphin, are distinguished by their long, narrow beaks and countershaded coloration that was light bluish gray on the back and pale gray to white on the belly. The long slightly upturned beak was used to probe the riverbed for food and given their small eyes and poor vision echolocation was employed to detect prey in murky water.

Baijis were most often observed in small groups of two to six individuals, but larger aggregations of up to 16 animals were sometimes found. Baijis were apparently opportunistic feeders eating a large variety of freshwater fish. They reached maturity at 4-6 years. Breeding took place in the first half of the year with most births occurring between February and April.

Historically, baijis occurred in the middle and lower reaches of the Chinese Yangtze River. Before the 1950s, the baiji was widely distributed in the Yangtze River. The first scientific surveys of the baiji were conducted in the late 1970s, and the population was already threatened by numerous human activities including fishing bycatch, habitat degradation due to dam construction, and heavy river traffic. Although laws were put in place banning harmful fishing practices regulations were not followed. The establishment of reserves in sections of the

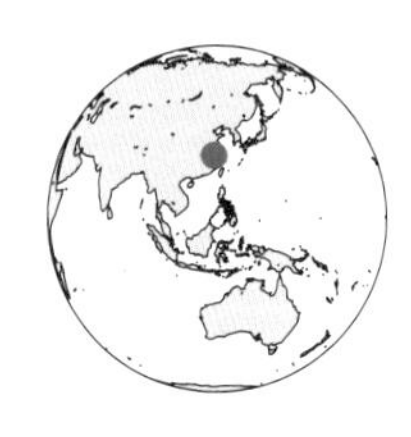

SIZE

Males: length up to 7 ft 6 in (2.3 m); weight 302 lb (157 kg)
Females: length up to 8 ft 6 in (2.6 m); weight over 368 lb (167 kg)

DIET

Freshwater fish

HABITAT

Fresh water, Yangtze River drainage

LIFE HISTORY

Sexual maturity
Males: 4 years
Females: 6 years
Gestation: 12 months
Reproduce: 2 years

STATUS

Likely extinct
Threats from fishing gear entanglement, overfishing, pollution, dam construction

Yangtze channel did not provide adequate protection. By the 1980s the population had been reduced to a few hundred and by the late 1990s surveys estimated around 13 individuals remained. Some attempts were made to establish a captive population, but the baiji's demise could not be halted and the last captive animal died in 2002. In 2006 a visual and acoustic survey failed to locate a single baiji and in 2007 the baiji was declared, in all likelihood, extinct.

A research team sequenced the genome of the baiji and found evidence for a genetic bottleneck that reduced species diversity 10,000 years ago and coincided with a decrease in temperature and rise of sea level. Sea level rise would have drowned the Yangtze River valley and led to a decline in available freshwater habitats. The extinction of the baiji, however, was not the result of this bottleneck but instead was caused by extreme human impacts to the Yangtze River in recent decades including pollution, harmful fishing gear, overfishing, and habitat destruction. Baijis also had low rates of reproduction giving birth every two years.

The baiji is the only member in the family Lipotidae and it was most closely related to the Amazon river dolphin and franciscana.

ABOVE
The Yangtze river dolphin resembled other river dolphin species in having tiny eyes and a long, slightly upturned beak.

OPPOSITE
The extinct Yangtze river dolphin(*Lipotes vexillifer*) was located in the Yangtze River. Its extinction was caused by human impacts, such as overfishing, pollution, and the construction of dams.

# NORTHERN SEA LION

## The northern sea lion is the largest of all sea lions. The western population underwent a marked decline in numbers in the 1980s and 1990s.

The northern or Steller sea lion, *Eumetopias jubatus*, exhibits significant sexual dimorphism. Adult males are larger and robust with a very thick neck, a blunt snout, and a lion-like mane. Adults are pale yellow to light tan in coloration. Males also have larger, thicker canines than females as well as thicker, wider chests, and shoulders. They also have a thick mane of fur extending from the back of the head to the shoulders.

Steller sea lions are polygynous with males aggressively obtaining and defending territories during the breeding season. Large groups at sea usually consist of females and subadult males. On land all ages and both sexes occur in loose aggregations during the nonbreeding season. Some Steller sea lions especially juveniles and adult males disperse widely outside the breeding season. Adults forage near their birth colonies and return to these sites to breed.

Steller sea lion males and females reach sexual maturity at about the same time between 3-7 years although males do not become territorial until they reach 9-13 years.

The breeding season lasts from mid-May to mid-July. Dominant males arrive early to establish territories on breeding rookeries on exposed rocks and beaches. Fighting among bulls involves chest to chest pushing matches but can escalate into bloody fights. Females give birth to a single pup three days after arriving at the rookery. About nine days after they give birth females begin a feeding cycle alternating between 1-3 days feeding at sea and 1-2 days nursing her pup. Closer to weaning the mother spends more time feeding at sea. The average nursing period is one year. Breeding season is followed by molting season. Females molt in early fall followed by males in late fall. Juveniles and nonbreeding Steller sea lions molt in late summer (July and August).

Steller sea lions are fast swimmers and can reach speeds of 18 mph (29 kph). Like all otariids they employ their foreflippers to produce propulsion and are able to "walk" on land. Diving is usually 656 ft (200 m) or less and dive duration is usually only a few minutes.

Steller sea lions do not bark; instead they produce low vocalizations, a deep roar, moving their head up and down while vocalizing. Adult males produce a range of vocalizations as part of their territorial behaviors and their primary function is social communication.

SIZE

Males: length 11 ft (3.4 m); weight 11 tons (10 tonnes)
Females: length 8 ft (2.5 m); weight 602 lb (273 kg)
At birth: length 3 ft (1 m); weight 35 lb–49 lb (16 kg–22 kg)

DIET

Fish (flounder, turbot, sardines) and invertebrates

HABITAT

Continental shelf and at sea

LIFE HISTORY

Sexual maturity
Males: 9–10 years
Females: 4–6 years
Gestation: 10 months
Reproduce: 1–2 years

STATUS

Near threatened
Endangered for western subspecies

An adult male northern sea lion (center) with females shows clearly the difference in size between the sexes as well as the lion-like mane of the male.

The massive decline (to approximately 20 percent of levels 40 years ago) in the 1970s and 1980s in one of two subspecies, the western northern sea lion, *E. j. jubatus,* that occurs west of Prince Edward Sound from Alaska to the Aleutians, has been attributed to a decline in quality and quantity of prey. Changes in the North Pacific ecosystem, likely due to both overfishing and global warming, have altered the availability of nutrient-rich salmon and herring. Instead, the sea lions are eating less nutritious cod and pollock, so-called "junk food." The junk food hypothesis has been tested and results indicate that although adults could survive eating pollock, yearlings could not eat enough to sustain them. Other scientists argue that other causes are responsible for sea lion decline and are the subject of ongoing research, such as changes in killer whale predation and commercial fishing of prey species. According to the predation hypothesis, killer whales that formerly hunted large whales' prey switched to other marine mammals, including the northern sea lion.

More recent review of the behavioral ecology of adult and juvenile Steller sea lions beginning in the 1990s supports the junk food hypothesis and suggests that plasticity in age at weaning might be key to differential survival of pups. Lactating females that consume large amounts of low-energy fish have a high probability of miscarriage and will keep their dependent young for an extra year or two, thereby causing declining birth rates and populations. In contrast lactating females that consume more fatty fish can successfully wean a pup a year. The eastern subspecies, Loughlin's Steller sea lion, *E. j. monteriensis,* that ranges from California to Prince Edward Sound, Alaska, has undergone a steady increase in numbers since the late 1970s.

Steller sea lions are members of the family Otariidae. They are most closely related to the California sea lion.

# Gray Whale

## Once common throughout the Northern Hemisphere, gray whales now only regularly inhabit the North Pacific Ocean.

Gray whales have a large, robust body and they are brownish-gray to light gray in color. The mouth is slightly arched and there are two to seven short, deep throat grooves used in suction feeding. They are covered with light colored patches of whale lice and barnacles, especially on the head and tail. The flippers are broad and paddle-like. A dorsal fin is absent but there is a dorsal hump and a series of "knuckles" along the dorsal ridge of the tail stock.

Of two extant Pacific populations, the eastern population, occurring in the North Pacific and Arctic Ocean, was once listed as endangered but successfully recovered and was delisted in 1994. The western population occurs in the western North Pacific off the eastern Asian coast. Although thought to be extinct in the Atlantic Ocean, there are rare reports of gray whales having been sighted in the Mediterranean Sea. Climate warming may be responsible for a change in migration of a few individuals following an Arctic route to the Atlantic. Whether this means a recolonization of gray whales in the Atlantic Ocean remains to be seen. But with an increase in global temperature and a reduction of ice in the Northwest Passage between the Pacific and Atlantic Oceans, it is a possibility. The likelihood of this is strengthened by study of the evolutionary relationships of Atlantic and Pacific populations of gray whales. Results indicate that the Atlantic populations are made up of four lineages, which are more closely related to different lineages in the Pacific populations than to each other. This means that each gray whale colonization event from the Pacific to the Atlantic resulted in the establishment of an Atlantic population that survived for millions of years, only to go extinct, be recolonized, go extinct, and so on.

Gray whales undertake one of the longest single migrations of any animal with a round trip of 9,000–13,000 miles (15,000–21,000 km). In the fall, Eastern North Pacific whales head south along the coast of North America to spend the winter in the warmer southern waters of California and Baja, and to mate and give birth. They do not form long lasting associations, but frequently travel alone or in small groups. Larger aggregations are observed on the feeding and breeding grounds. Over the last several years, fewer gray whales are arriving in Mexican lagoons, and many are malnourished: some even wash up dead on the shore. Even more concerning is the dramatic drop in births. Between 2016 and 2020 the

SIZE

Males: length 36 ft–49 ft (11 m–15 m), females are slightly larger;
Both sexes: weight up to 49½ tons (45 tonnes)
At birth: length 15 ft–16 ft (4.6 m–4.9 m); weight 2,028 lb (920 kg)

DIET

Mysids, amphipods, polychaete tube worms

HABITAT

Shallow, coastal waters, migratory

LIFE HISTORY

Sexual maturity
Males & females: 9 years
Gestation: 11–13 months
Reproduce: every 2 years

STATUS

Least concern
Critically endangered western Pacific stock

A breaching adult gray whale reveals prominent patches of whale lice on the skin, a useful identification feature.

estimated population of gray whales plummeted by nearly a quarter from 27,000 to just over 20,000. Although the reason for the decline is not known, climate change is a likely major factor causing a reduction of the quality and quantity of their food supply.

Gray whales are lateral suction feeders rolling onto their sides to feed on benthic invertebrates, mostly amphipods, from the seafloor. They often leave long feeding trails of mud. A study by paleontologists Nick Pyenson and David Lindberg suggests that gray whales may have had different feeding strategies and food in the past. They showed that changes in Pleistocene sea level changes resulted in fluctuations of available benthic feeding areas. According to these scientists, gray whales survived periods of maximum ice by feeding outside the Bering Sea and Sea of Okhotsk, which were ice covered. Gray whales likely employed more generalized feeding strategies that enabled them to feed on alternative prey, such as fish, similar to the resident gray whales found today off the Pacific Northwest coast.

Gray whales are the only members of the baleen whale family Eschrictiidae.

# Antarctic Minke Whale

## One of only two species of minke whale, Antarctic minke whales are also known as southern minke whales. They are affected by climate change.

Antarctic minke whales (*Balaenoptera bonaerensis*) are among the smallest of rorqual species. Rorquals are members of the family Balaenopteridae, meaning furrow whale in Norwegian, so-named because of their extensive throat grooves, extending from under the mouth to the belly. Females are slightly larger than males. They are dark gray on the back with a pale belly. They can be distinguished from their close relative common minkes by the lack of a white patch on their flippers. Initially they were considered a subspecies of the common minke whale. Since the 1990s they have been recognized as a distinct species.

Antarctic minke whales are circumpolar and occur widely in coastal and offshore areas of the Southern Hemisphere. The species is associated with sea ice, especially around the ice edge, and is generally less abundant in ice-free waters. Their small, compact bodies and short fins make them well suited to living in pack ice where they can maneuver in narrow spaces between ice floes. Although not all whales migrate, there is a general shift northward for breeding in the winter months. Known breeding grounds are off Australia, South Africa, and Brazil. Like some other baleen whales, some Antarctic minke whales remain in the Antarctic all year. In the Antarctic, they co-occur with the common minke whale and hybrids of the two species have been reported.

In feeding areas, minkes can be solitary or form small groups of two to four individuals; larger feeding aggregations of several hundreds of animals can also occur. They feed primarily on Antarctic krill, although they also consume various species of fish. Monitoring of depth, acceleration, and body orientation using digital tags on the whales revealed that the minkes employed a unique hunting strategy, skimming the underside of the ice and rapidly gulping krill, making then the fastest "lunge feeders" known. The minkes lunge up to 24 times during a single dive, nearly once every 30 seconds. Scientists suggest that this near-surface, multi-gulp strategy may be well suited to consuming patches of krill. Antarctic minke whales are highly associated with sea ice and the reduction in sea ice due to climate warming decreases the abundance of Antarctic krill, threatening the survival of minkes and other Antarctic predators relying on krill availability.

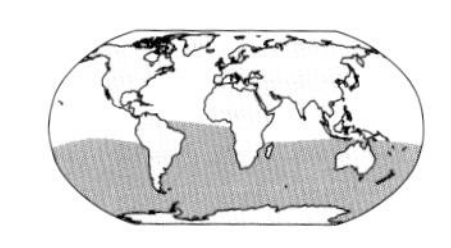

SIZE

Males & females
Length: 28 ft 10 in–35 ft (8.5–10.7 m)
Weight: 6.85 tons–11.05 tons (6.2 tonnes–10 tonnes)
At birth: length 9 ft (2.8 m); weight: 700 lb–1,000 lb (317 kg–453 kg)

DIET

Mostly krill, occasionally schooling fish

HABITAT

Coastal and offshore waters

LIFE HISTORY

Sexual maturity
Males: 8 years
Females: 7.8 years
Gestation: 10 months
Reproduce: yearly

STATUS

Data deficient
Threat from hunting by humans

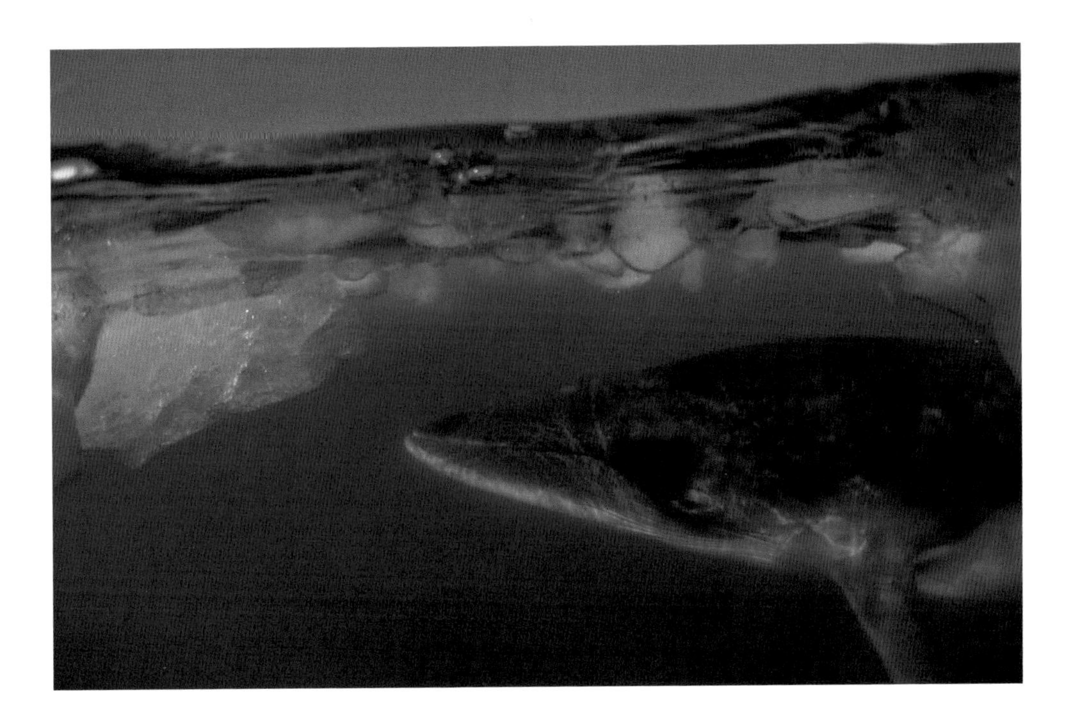

ABOVE
Antarctic minke whale diving at the Lemaire Channel, Antarctica. This whale has a coating of yellowish diatoms on the skin. It has been suggested that this species and others that live in cold waters migrate to warmer waters to shed this harmful coating.

OVERLEAF
A pod of Antarctic minke whales in front of icebergs at the Gerlache Strait, Antarctic Peninsula, Antarctica. The short, bushy blow is a conspicuous identification feature.

Antarctic minke whales are fast swimmers reaching 20 mph (32 kph). They occasionally exhibit aerial maneuvers such as breaching, leaping nearly completely out of the water, or porpoising—low leaps made at the water's surface.

These whales produce a variety of sounds, including whistles, calls, clicks, screeches, grunts, and down sweeps (a signal that decreases in frequency over time). They also produce a unique vocalization called the "bioduck sound," which is a duck-like quacking sound that had mystified scientists for many years.

Although initially not hunted during the commercial whaling of the nineteenth century because of their small size and relatively fast movements, once the larger whales were depleted, Antarctica minke whales were hunted in the early 1970s. Whaling continued in Japan in the form of "scientific whaling" conducted under the International Whaling Commission (IWC) approved program. In 2019, Japan left the IWC and no longer commercially hunts whales in Antarctica. However, this species is now at risk due to climate change, which has caused a reduction of sea ice, which this species relies on for foraging, habitat, and prey. Other threats are ship strikes caused by the advent of larger, faster moving ships, as well as military and seismic noise underwater explosions, and entanglements in fishing nets.

Antarctic minke whales and the common minke whale are members of the baleen whale family Balaenopteridae, and they are most closely related to gray whales (Eschrictiidae).

# Harp Seal

## Harp seals have been commercially hunted for oil, pelts, and meat, and although sealing activities have declined, they continue to be hunted today in Canada.

The most common Antarctic pinniped, the harp seal, *Pagophilus groenlandicus*, means "the ice lover from Greenland." The most distinctive feature of harp seals is their coloration pattern that changes over their lifespan. Newborns have a pure white pelage that persists for about 12 days during which time they are known as "whitecoats." The white fur helps absorb the sunlight and trap heat to keep the pups warm. As they age, various spotted stages follow that develop into a dark "harp" pattern (the basis of their common name) on their back and sides, and the spots disappear.

Harp seals exhibit little sexual dimorphism; males are slightly larger than females. Vision is their primary sense. They have well-developed beaded whiskers that are likely used to detect the movement of fish and other aquatic organisms. They are generally shallow divers foraging at 328 feet (100 m) or less. They are known for their varied vocalizations during the breeding season. Female harp seals use their sense of smell to identify their own offspring during the nursing period.

Harp seals are widespread in the Arctic and North Atlantic Oceans from northern Russia in the east to eastern Canada and the northeastern US in the west. They spend most of their time near pack ice, especially during breeding and molting seasons.

Harp seals are gregarious and congregate on pack ice where they form huge groups sometimes numbering into the thousands. To attract females, some male harp seals blow bubbles and vocalize below the ice near where females have made entry holes in the water. Males also may chase females. To compete for females, males may splash and bite other males. They are highly migratory and after the breeding season and molt, they follow the ice north in summer to feed in the Arctic. They feed on a wide array of fish and crustaceans and their feeding habits vary with age, season, location, and year.

Harp seals are members of the family Phocidae most closely related to the ribbon seal.

SIZE

Males: length up to 6 ft (1.9 m); weight 265 lb–298 lb (120 kg–135 kg)
Females: length up to 6 ft (1.8 m); weight 265 lb (120 kg)
At birth: length 2 ft 7 in (0.8 m); weight 9 lb (4 kg)

DIET

Crustaceans and fish (capelin, cod)

HABITAT

Arctic and North Atlantic oceans

LIFE HISTORY

Sexual maturity
Males: 7–8 years
Females: 4–7 years
Gestation: 11.5 months
Reproduce: yearly

STATUS

Least concern

An adult harp seal showing the distinctive "harp" pattern on its back and its dark head.

## WHY DOES SEAL HUNTING CONTINUE?

Archaeological evidence indicates that indigenous people in the US and Canada have been hunting seals for at least 4,000 years. Newfoundland and Labrador were the first regions to experience large-scale seal hunting or sealing in the early eighteenth century and the meat, fur, and oil of the seals were utilized. Initially small vessels were used that were replaced by steamships in the mid-eighteenth century. By the end of the nineteenth century various laws enacted curtailed most commercial sealing.

The annual hunt for harp seals in Canada is the largest commercial take of marine mammals. Only adults are hunted and since 1987 the hunting of pups—"whitecoats"—has been illegal. In the mid-2000s, seal hunters regularly killed around 330,000 seals per year. In 2009, a ban on the trade of seal products by the European Union significantly reduced this number. Less than 400 seals were killed in 2020 in Canada, which was only 8 percent of the 2019 quota of 400,000 seals. Today, the Canadian seal industry is greatly diminished. One argument fishermen use to justify their hunting is that the seals are feeding on an economically important fish—cod—and hurting the fishing industry. Scientists counter that cod makes up only a small portion of a seal's diet and the fishing industry, not seals, is to blame for the decline. Despite the pressure of animal welfare groups, sealing continues not just in Canada but also in Namibia (Cape fur seals) and Greenland (hooded, ringed, and harp seals).

# HECTOR'S DOLPHIN

## Historically, Hector's dolphins were victims of entanglements in gillnets. Species recovery has largely been the result of their protection at the Banks Peninsula Marine Sanctuary.

Hector's dolphins (*Cephalorhynchus hectori*), endemic to New Zealand, are among the smallest dolphins. Females are larger than males. Two subspecies are recognized, *C. h. hectori* and *C. h. maui*, based on morphological and genetic differences. Maui's dolphin is rarer and found only along the west coast of the North Island, whereas Hector's dolphin is mainly found around the South Island. It is slightly longer than the South Island subspecies. A 2021 report issued by the New Zealand government indicates that the population of Maui's dolphin is around 50 individuals. There is no evidence that the two subspecies interbreed.

Hector's dolphins are light gray with the dorsal fin, flukes, flippers, and much of the face dark gray to black. They prefer shallow waters less than 328 feet (100 m) deep and 52 feet (16 m) from shore. Their inshore habitat puts them in direct contact with human activities including fishing, pollution, and marine mining among others. By far the most effective conservation management involves removing gillnet and trawl fisheries, which can be achieved by changing to more selective fishing methods that do not trap dolphins. Commercial gillnet fishing banned at Banks Marine Sanctuary has increased their population numbers.

Hector's dolphins are generalist, opportunistic feeders, and their diet includes cephalopods, crustaceans, and schooling small fish. The diet is more varied on the east coast than the west coast of the South Island; it is not known whether these differences are in prey preference or availability.

Hector's dolphins have low genetic diversity. Genetic analyses support differentiation of East and West Coast populations as well as two South Island populations. Likely deep water is the principal factor as it limits the distribution and density of these dolphins. There appears to be male-biased dispersal that suggests protection of dispersal corridors would be important. A patten of male-biased dispersal would be consistent with the suggestion that males are likely to encounter more nets while moving between areas, thereby increasing the likelihood of entanglement in nets.

Hector's dolphins live in small groups of two to eight individuals, although large groups of up to 50 are occasionally seen, the result of the coalescence of several small groups. Mother-calf pairs typically associate with other females. Maui's dolphins show a different pattern with mixed sex aggregations, perhaps a

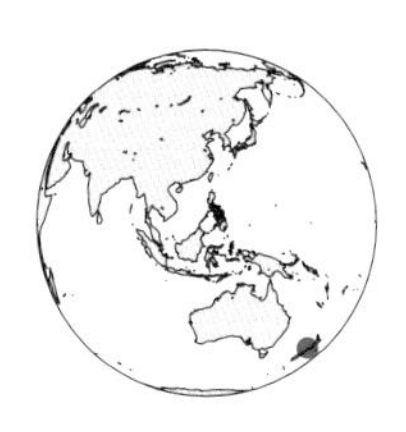

SIZE

Males: length 5 ft (1.5m); weight up to 143 lb (65 kg)
Females: length 5 ft (1.5 m); weight 110 lb (50 kg)
At birth: length 1 ft 11 in–2 ft 3 in (0.6–0.7 m); weight 20 lb (9 kg)

DIET

Small fish and squid

HABITAT

Shallow coastal waters

LIFE HISTORY

Sexual maturity
Males & females: 5–9 years
Gestation: 10–11 months
Reproduce: 2–4 years

STATUS

Critically endangered (Maui subspecies)
Endangered
Threat from fishing gear

Hector's dolphins showing the characteristic dark coloration of the face, dorsal fin, flukes, and flippers.

consequence of their extremely small population size. Similar to other dolphins they have a fission-fusion social system, one in which the size and composition changes through time. They are active and sometimes acrobatic, leaping out of the water and engaging in bow riding.

Male Hector's dolphins have large testes compared to body size, which suggests that males compete for access to females. Females don't have their first calf until they are 7–8 years old and they reproduce only every 2–4 years; both these factors coupled with their low genetic diversity is detrimental to conservation efforts and may result in low survival rates.

Hector's dolphins are in the family Delphinidae and they are most closely related to Commerson's and Chilean dolphins.

# Sea Mink

## The sea mink, best known from Native shell middens along the coastal islands of the Gulf of Maine, is the only mustelid hunted to extinction.

Except for otters, the sea mink (*Neovison macrodon*) is likely the most aquatic of mustelids and provides a late Cenozoic example of a species that recently and rapidly became marine. The sea mink can be distinguished from the American mink by its larger overall size and robustness. Historic accounts describe them as "extremely fat." Its large size contributed to its desirability to fur traders. When the sea mink was first described in 1903, it was already extinct. It is known from hundreds of individual bones found in archaeological shell middens. Its distribution was likely along the rocky coasts of New England and the Atlantic Canada as far north as Nova Scotia although it may have ranged further south during the last glacial period. It likely lived from 5,000 to 150 years ago.

The body of the sea mink was flatter than the American mink and it had a long bony tail and a coarse reddish-brown coat. Females were smaller than males. Sea minks had blunter and wider teeth than American mink. Fish bones found in likely den sites suggest that sea minks fed mainly on aquatic prey but they also may have preyed on seabirds, seabird eggs, and hard-shelled invertebrates. It is possible that the sea mink was an important predator in the coastal and intertidal ecosystems of the North Atlantic.

Although fur trade hunting was the most likely cause of extinction other contributors include indigenous hunting and extinction of the Labrador duck with which it coexisted and which may have been a prey item. After the Civil War, pelt prices astronomically increased and sea minks became rarer, requiring dogs to find them in their burrows and flush them out along the rocky intertidal.

The sea mink is a member of the family Mustelidae and it is most closely related to the American mink.

SIZE

Males & females
Length: average
3 ft (0.9 m)
Weight: unknown

DIET

Aquatic species, including fish

HABITAT

Coastal waters

LIFE HISTORY

Sexual maturity: Unknown
Gestation: unknown
Reproducing: unknown

STATUS

Extinct in historic times

Sea mink were distinguished from their close relative the American mink by their larger size, more robust body, and longer tail.

# POLAR BEAR

## The polar bear is found throughout the Arctic. Loss of sea ice due to climate change has reduced their numbers and continues to threaten their future.

Polar bears (*Ursus maritimus*), the largest and most carnivorous of eight species of bears, called the "poster child of climate change" may become extinct by the end of the century if stricter climate policies are not enacted. A 20-year study of polar bears in Svalbard, Norway, found a 10 percent loss in genetic diversity. Scientists concluded that climate change and the rapid loss of arctic ice were causing habitat fragmentation. Studies of polar bears in Baffin Bay have revealed that due to climate changes in the Arctic over the last 25 years polar bears were getting thinner and having fewer cubs. Both are linked to sea ice availability since when they are on land they don't hunt seals; instead they rely on fat reserves.

As many as 20 distinctive polar bear populations have been identified. One isolated population that diverged from other populations just a few hundred years ago lives in southeast Greenland. These genetically distinct bears have developed a previously unrecognized behavior. They use fresh ice at the front of glaciers as a platform to hunt seals. Scientists further suggest this behavior has implications for conservation since it has aided the survival of polar bears in this new habitat and mitigated the detrimental effects of sea ice decline.

Generally, the coat of polar bears is white but depending on lighting it can appear yellow, light brown, or even light gray. Compared to other bears, polar bears have smaller ears, a slender neck and head, and short, sharp claws that provide traction on the ice. The forepaws are large and oar-like—adaptations for swimming and walking on ice. Males are larger than females.

The principal prey of polar bears are seals associated with ice, mainly ringed seals. They typically aggregate in areas of high ringed seal density, where polar bears hunt by waiting near a breathing hole in the ice used by the seals. When the seal surfaces, the polar bear pulls it onto the ice and devours it. Study of the dietary behavior of polar bears by examining tooth wear shows that they became highly specialized feeders consuming the blubber and flesh of seals after they diverged from brown bears. Warming climate and the changing landscape mean that brown bears can venture further north and compete with polar bears for available food and given their highly specialized diet the survival of polar bears is not favored.

SIZE

Males: length up to 8 ft (2.5 m); weight up to 1,764 lb (800 kg)
Females: length 6 ft 6 in (2 m); weight 772 lb (350 kg)

DIET

Primarily ringed seals

HABITAT

Arctic sea ice

LIFE HISTORY

Sexual maturity
Males: 6 years
Females 4 years
Gestation: 8 months
Reproduce: 2–4 years

STATUS

Vulnerable
(some sub-populations are data deficient)
Threats from human impact, climate change

ABOVE
A swimming polar bear using its large oar-like forelimbs.

OVERLEAF
Polar bears spend most of their time on ice and its loss in a warming world threatens their survival.

However, a previously unidentified population of polar bears from the southeastern coast of Greenland survive on glacial ice and display foraging strategies that may allow their survival in a warming climate.

Polar bears tend to be solitary but breeding pairs and mothers with cubs may be seen together. They spend most of their time on ice, but sometimes spend considerable time on land. Only pregnant polar bears go into dens for the winter, while other bears remain on the ice and hunt throughout the winter. Although polar bears grow up to be huge, the cubs are born blind, deaf, lightly furred, and weigh only about 1 lb (0.5 kg). Cubs stay sheltered in snow dens during the first months of their life, depending completely on their mothers for food, warmth, and protection. Key aspects of their denning behavior has been documented using remote solar-powered cameras.

Polar bears make long migrations of 1,243–2,485 miles (2,000–4,000 km) across the ice in response to ice packs receding in the spring and advancing in the fall. In addition, they may travel large distances to find mates or food.

Although today hunting is not the major threat to polar bears, they were hunted in the past and are taken in small numbers today by the indigenous peoples of Alaska, Canada, Greenland, and Siberia.

Polar bears are most closely related to brown bears, diverging from them between 5–4 mya to less than 600,000 years ago. Polar bears are members of the family Ursidae.

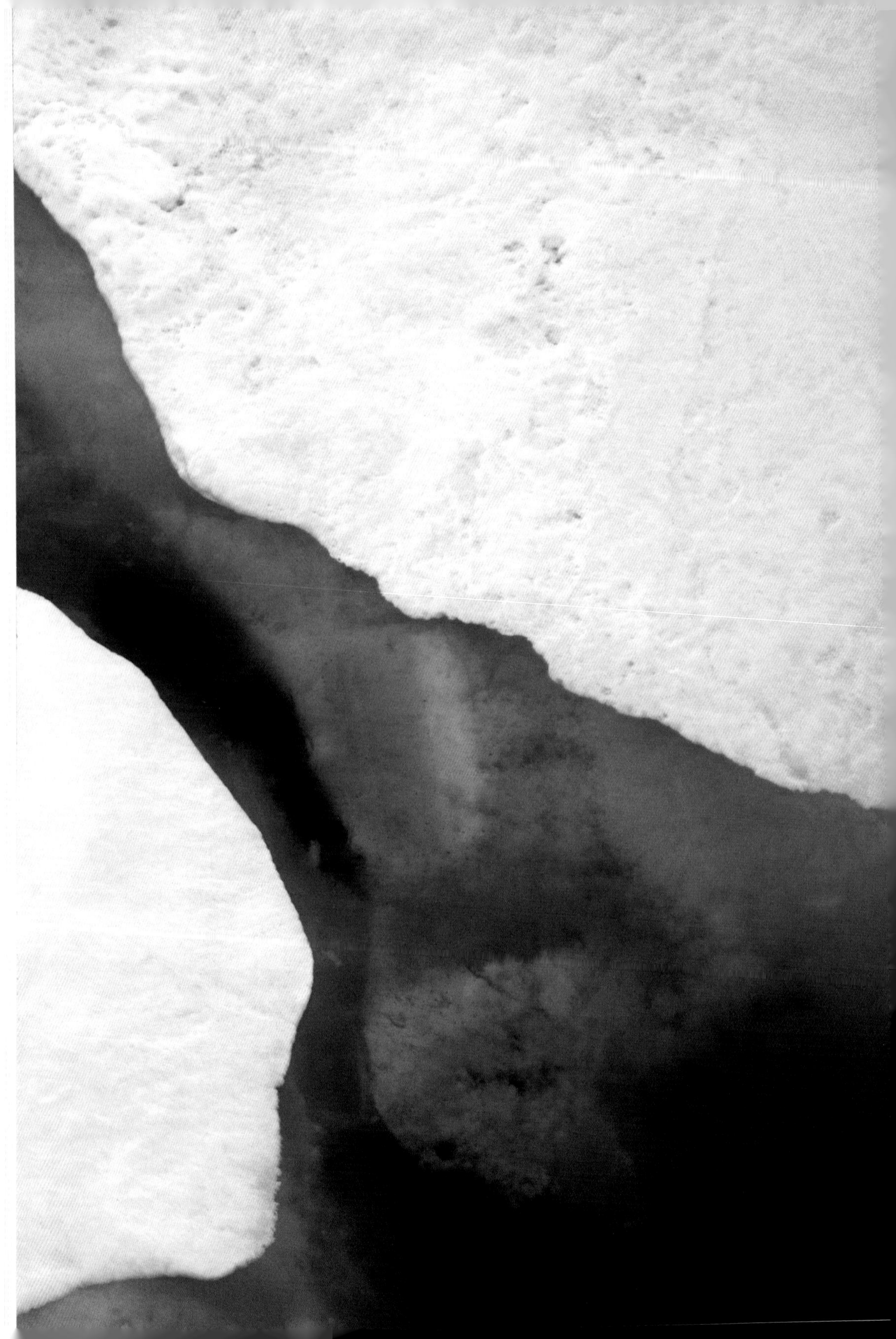

# Picture Credits

The publisher would like to thank the following for permission to reproduce copyright material:

Cover
Flip Nicklin/naturepl.com

Agefotostock
Auscape/UIG 66b; Konstantin Mikhailov/imageBROKER 122b; Bruce Watkins 191

Alamy Stock Photo
Marko Steffensen 55; Florilegius 66t; Eraza Collection 71; WILDLIFE GmbH 106; Nikolay Vinokurov 122t; imageBROKER 124, 151; Natalia Pryanishnikova 133; Robert Bannister 147; Eric Wengert 153; Mark Hicken 166; Design Pics Inc 179b, 181b; Michele and Tom Grimm 189t

Ardea Picture Library
M. Watson 107; ardea.com/Samuel Blanc/Biosphoto 119b; Nick Gordon 194; Keith & Liz Laidler 195

Blue Planet Archive
D. R. Schrichte 13; John K. B. Ford/ Ursus 59b; Masa Ushioda 65; Scott Hanson 102; Steven Kazlowski 112; Kevin Schafer 115, 205; Doug Allan 138b; Doug Perrine 155, 161-163, 175, 207; Neil Ever Osborne 138; D.R. Schrichte 159; Jon Cornforth 179t; John Gibbens 187; Christopher Swann 199

Florida Fish and Wildlife Conservation Commission, taken under NOAA research permit #15488 185

Florida Museum of Natural History 7

Getty Images
Southern Metropolis Daily 48; VW Pics 95; Paul A. Souders 101; The Asahi Shimbun Premium 119t; Mike Korostelev 141; James R.D. Scott 148t; Sven Gruse/EyeEm 148b; Adam Cropp 201; Kerstin Langenberger 202-203; by wildestanimal 144-145

©Thomas A. Jefferson/VIVA Vaquita 171, 173t

Nature in Stock
Flip Nicklin/Minden Pictures 59t, 62-63, 91, 103, 113; Tui De Roy/ Minden Pictures 165; Sebastian Kennerknecht/Minden Pictures 181t; Jon Baldur Hlidberg/Wildlife/Minden Pictures 183

Naturepl.com
Enrique Lopez-Tapia 9; Eric Baccega 61,197; Brandon Cole 84, 116b; Alex Mustard 89, 156-157; Doc White 90; Espen Bergersen 94; Todd Mintz 97; Tony Wu 98-99, 130; Sylvain Cordier 105t, 116t; Sergey Gorshkov 109; Andy Rouse 110-111; Olga Kamenskaya 121t; Klein & Hubert 129; Roland Seitre 134; Hugh Pearson 135; Stefan Christmann 138t; Luis Quinta 142; Wild Wonders of Europe/Sà 177; Piper Mackay 189b; Todd Pusser 183; Steven Kazlowski 211-213

Eric Angel Ramos, Fundación Internacional para la Naturaleza y la Sustentabilidad 131

©www.savethewhales.org 173b

Science Photo Library
Cecile Degremont/Look at Sciences 14; Chris & Monique Fallows/Nature Picture Library 93

Shutterstock
Ranjan Barthakur 10; Andrei Gilbert 121b; Katiekk 137; spline_x 143 Wonderful Nature 146; Sylvie Bouchard 211

Miniature scale 3D print of fossil whale specimen Museo Paleontologico de Caldera (MPC) 675, Courtesy Smithsonian Institution National Museum of Natural History, Department of Paleobiology 82t

3D model of fossil whale specimens Museo Paleontologico de Caldera (MPC) 665-667, Courtesy Smithsonian Institution National Museum of Natural History, Department of Paleobiology 82b

Image reproduced courtesy of The Wellcome Library, London 60

Hans Wolkers 105b

All reasonable efforts have been made to trace copyright holders and to obtain their permission for use of copyright material. The publisher apologizes for any errors or omissions in the list above and will gratefully incorporate any corrections in future reprints if notified.

# ACKNOWLEDGMENTS

First and foremost, thanks to the very knowledgeable and professional editorial team at UniPress, especially commissioning editor Kate Shanahan, who facilitated project development, and designer Alexandre Coco who provided excellent looking pages. Bob Nicholls skillfully restored extinct sea mammals, bringing the past back to life, and Alison Stevens assisted by researching stunning photographs of sea mammals in action. Special thanks to my editor and project manager Kate Duffy for her patience throughout the two-year production process, especially her thorough and exacting copy editing. Finally, I am grateful to numerous colleagues and graduate students for providing inspiration, enthusiasm, and guidance.